Semi-Riemannian Maps and Their Applications

Mathematics and Its Applications

Volume 475

Semi-Riemannian Maps and Their Applications

by

Eduardo García-Río
University of Santiago de Compostela, Santiago, Spain

and

Demir N. Kupeli
Middle East Technical University, Ankara, Turkey

SPRINGER-SCIENCE+BUSINESS MEDIA, B.V.

A C.I.P. Catalogue record for this book is available from the Library of Congress.

ISBN 978-90-481-5202-5 ISBN 978-94-017-2979-6 (eBook)
DOI 10.1007/978-94-017-2979-6

Printed on acid-free paper

To Professor Arthur E. Fischer,
through whose inspiration
the main ideas
in this book developed

Contents

Preface

A major flaw in semi-Riemannian geometry (as compared to other subjects) is a shortage of suitable types of maps between semi-Riemannian manifolds that will compare their geometric properties. A class of such maps is introduced in this book and is called semi-Riemannian maps. Roughly speaking, a map between semi-Riemannian manifolds is semi-Riemannian if it is "as isometric as it can be". Consequently, being generalizations of isometries, these maps will compare geometric properties of semi-Riemannian manifolds.

A semi-Riemannian map between Riemannian manifolds is called a Riemannian map. The concept of a Riemannian map was first introduced by Arthur E. Fischer [F1] as a generalization and unification of the notion of isometric immersion and Riemannian submersion. He showed that a Riemannian map is automatically a subimmersion, and thus a Riemannian subimmersion. If the notion of isometric immersion and Riemannian submersion is considered to be the appropriate Riemannian generalization of the notions of immersion and submersion that occur in the theory of maps between manifolds, then the notion of a Riemannian map may be considered to be the appropriate Riemannian generalization of the notion of subimmersion that occurs in the theory of maps between manifolds. But a semi-Riemannian map between semi-Riemannian manifolds is not necessarily a subimmersion, so it cannot be considered the sole analog of a Riemannian map in semi-Riemannian geometry. In fact, a semi-Riemannian map between semi-Riemannian manifolds possesses a certain semi-Riemannian nature which makes it more interesting in semi-Riemannian geometry.

The main purpose of this book is to state some results in semi-Riemannian geometry obtained by the existence of a semi-Riemannian map between semi-Riemannian manifolds, as well as to encourage the reader to explore semi-Riemannian maps. It seems semi-Riemannian maps deserve further investigation and they may become a research area like harmonic maps.

In Chapters 1, 2 and 3, we develop fundamental concepts and formulas in semi-Riemannian geometry which will be used throughout the book. Even the reader familiar to semi-Riemannian geometry is strongly recommended to read these chapters since they contain certain material which is not available in the standard literature in semi-Riemannian geometry. Especially in Section 1.2, we develop the main linear algebraic concepts behind semi-Riemannian maps.

In Chapter 4, we introduce the notion of semi-Riemannian map. Here we will make the first application of semi-Riemannian maps to semi-Riemannian submersions which admit degenerate submanifolds as fibers. Also we will give some attention to affine Riemannian maps which may be considered as "partial isometries".

In Chapter 5, we generalize semi-Riemannian maps and semi-Riemannian submersions to semi-Riemannian maps with respect to a semi-Riemannian foliation and to semi-Riemannian transversal maps, respectively. Here we also obtain the fundamental equation of a semi-Riemannian transversal map.

Chapter 6 is devoted to a special case of semi-Riemannian maps from a semi-Riemannian manifold to a 1-dimensional semi-Euclidean space. Here we also prove the semi-Riemannian Regular Interval Theorem which will be used in the proof of the splitting theorems in Chapter 7.

Finally in Chapter 7, we make some applications of semi-Riemannian maps to splitting theorems in semi-Riemannian, Riemannian and Lorentzian geometry. The splitting theorems in Lorentzian geometry may also be of interest to physicists since they may also be considered as rigid singularity theorems in General Relativity.

Appendices A and B are written to make the book more self-contained. In Appendix A, we review some fundamental facts about submanifolds of semi-Riemannian manifolds which are encountered in this book. In Appendix B, we review certain fundamental concepts in Riemannian and Lorentzian geometry which are used in this book.

The level of this book has been kept as elementary as possible so as to be accessible by graduate students who have some background knowledge in differentiable manifolds and elementary differential topology. The concepts of semi-Riemannian geometry that we use are developed in the book, and hence it may be considered as self-contained in this respect.

This work was supported through the grants DGICYT (TXT97-2075) and XUGA (20702B96) from Spain. The second-named author (DNK) is grateful to the Department of Geometry and Topology of the University of Santiago de Compostela for their kind invitation and for providing support while writing this book.

Finally, we wish to thank Luis A. Cordero for several discussions on transversal maps to a foliation.

November 1998

Eduardo García-Río

Demir N. Kupeli

Chapter 1

Linear Algebra of Indefinite Inner Product Spaces

In this chapter we collect the mechanisms behind the linear algebra of real inner product spaces. Some of these were developed for certain classes of geometric objects in other literature, yet their proofs involve only the linear algebraic properties of the geometric objects at hand.

In semi-Riemannian geometry, a nondegenerate bilinear form on a real vector space is called an inner product. However, unlike definite inner products, an indefinite inner product may be degenerate when restricted to certain subspaces. We call such subspaces degenerate. Degenerate subspaces have a certain importance in semi-Riemannian geometry and therefore we devote the first section of this chapter to some properties of degenerate subspaces. The second section of this chapter studies linear transformations between inner product spaces. We define the nondegenerate rank and the adjoint of a linear transformation which will be the essential parts of the main definitions throughout this book. In this section, we also introduce an inner product on the space of linear transformations between inner product spaces. The square norm induced by this inner product and the nondegenerate rank will then be related by semi-Euclidean linear transformations.

1.1 Inner Product Spaces

Throughout this section, let V denote an n-dimensional real vector space.

Let $g : V \times V \to \mathbf{R}$ be a bilinear form on V. It is well-known from linear algebra that there exists an ordered basis $\mathcal{B} = \{\alpha_1, \ldots, \alpha_n\}$ for V such that,

(1) $g(\alpha_i, \alpha_j) = 0$ for $i \neq j$,

(2) $g(\alpha_i, \alpha_i) = 0$ for $1 \leq i \leq \mu$,

(3) $g(\alpha_i, \alpha_i) = -1$ for $\mu + 1 \leq i \leq \mu + \nu$,

(4) $g(\alpha_i, \alpha_i) = 1$ for $\mu + \nu + 1 \leq i \leq \mu + \nu + \eta$,

where μ, ν, η are certain integers. Also (μ, ν, η) is called the *type* of g and the basis $\mathcal{B}$ is called an *orthonormal basis for V with respect to g* (cf. [No, p. 107], [HK, p. 370]).

Definition 1.1.1 Let g be a bilinear form on V. The *degenerate space* L_g of g is defined by

$$L_g = \{x \in V / g(x, y) = 0 \text{ for all } y \in V\}.$$

Remark 1.1.1 Let g be a bilinear form of type (μ, ν, η) on V. Then first note that L_g is a subspace of V and dim $L_g = \mu$. Also $\nu = \sup\{$ dim W / W is a subspace of V with $g(x, x) < 0$ for all $x \in W\}$ and $\eta = \sup\{$ dim W / W is a subspace of V with $g(x, x) > 0$ for all $x \in W\}$. (See [HK, p. 370]).

Definition 1.1.2 Let g be a bilinear form of type (μ, ν, η) on V. If $\mu = 0$ (that is, if g is *nondegenerate*) then g is called an *inner product* (*of index* ν) on V and (V, g) is called an *inner product space* (*of index* ν). Also, if (V, g) is an inner product space (of index ν) then an orthonormal basis $\mathcal{B}$ for V with respect to g is called an *orthonormal basis* for (V, g). Also, an inner product space (V, g) of index $\nu = 0$ is called a *positive definite inner product space.*

For sake of brevity in terminology, we will also assign the following causal characters to the vectors in an inner product space.

Definition 1.1.3 Let (V, g) be an inner product space. A nonzero vector $v \in V$ is called

(1) *timelike* if $g(v, v) < 0$,

(2) *null* if $g(v, v) = 0$,

(3) *spacelike* if $g(v, v) > 0$,

(4) *causal* if $g(v, v) \leq 0$.

Let (V, g) be an inner product space of index ν. If $\nu = 0$ or $\nu = n$ then the restrictions of g to the subspaces of V are inner products on these subspaces of index $\nu = 0$ or $\nu = n$, respectively. However if $0 < \nu < n$ then the restrictions of g to some subspaces of V may be bilinear forms of type (μ', ν', η') with $\mu' \geq 1$. This motivates the definition of degenerate subspaces.

Definition 1.1.4 Let (V, g) be an inner product space. Let W be a subspace of V and g_W be the restriction of g to W. If g_W is an inner product on W (of index ν') then (W, g_W) is called a *nondegenerate subspace of* (V, g) (*of index* ν'). If g_W is a bilinear form of type (μ', ν', η') on W with $\mu' \geq 1$, then (W, g_W) is called a *degenerate subspace of* (V, g) (*of type* (μ', ν', η')).

Next we will investigate some properties of subspaces of inner product spaces. For any subspace W of an inner product space (V, g), we define the *orthogonal space* $W^\perp$ to W in (V, g) as usual by

$$W^\perp = \{x \in V / g(x, y) = 0 \ \text{for all}\ y \in W\}.$$

Proposition 1.1.1 *Let* (V, g) *be an inner product space and let* W *be a subspace of* (V, g). *Then*

(1) $\dim W + \dim W^\perp = \dim V = n$

(2) $(W^\perp)^\perp = W$

(3) (W, g_W) *is a nondegenerate subspace of* (V, g) *if and only if* $W \cap W^\perp = \{0\}$ *if and only if* $W \oplus W^\perp = V$.

Proof. (1) Let $\{\alpha_1, \dots, \alpha_k\}$ be a basis for W and let $\{\alpha_1, \dots, \alpha_n\}$ be a completion of $\{\alpha_1, \dots, \alpha_k\}$ to a basis for V. Then $x \in W^\perp$ if and only if $g(x, \alpha_i) = 0$ for $i = 1$, 2, ..., k. Let $W^\perp \ni x = \sum_{i=1}^n x^i \alpha_i$ and $g_{ij} = g(\alpha_i, \alpha_j)$. Hence $\sum_{i=1}^n x^i g_{ij} = 0$ for $i =$ 1, ..., k. But since $[g_{ij}]$ is nonsingular, the row rank of the coefficient matrix of the linear system

$$\sum_{i=1}^n x^i g_{ij} = 0, \qquad i = 1, \dots, k$$

is k and hence its solution space has dimension $n - k$. Thus $\dim W^\perp = \dim V - \dim W$.

(2) Since $W \subseteq (W^\perp)^\perp$ and $\dim W = \dim (W^\perp)^\perp$ by (1), it follows that $(W^\perp)^\perp = W$.

(3) First note that, if u is in the degenerate subspace L_{gw} of g_W in W, then u is orthogonal to every vector in W, and hence $u \in W^\perp$. Thus $L_{gw} \subseteq W \cap W^\perp$. Also if $u \in W \cap W^\perp$ then u is orthogonal to every vector in W, and hence $u \in L_{gw}$. Thus $W \cap W^\perp \subseteq L_{gw}$. Therefore $L_{gw} = W \cap W^\perp$. Hence it follows that (W, g_W) is nondegenerate if and only if $W \cap W^\perp = \{0\}$. For the final equivalence, since

$$\dim (W + W^\perp) + \dim (W \cap W^\perp) = \dim W + \dim W^\perp,$$

it follows that $V = W \oplus W^{\perp}$ if and only if $W \cap W^{\perp} = \{0\}$. □

Remark 1.1.2 Note that since $(W^{\perp})^{\perp} = W$, it also follows that $L_{g_W} = W \cap W^{\perp} = L_{g_{W^{\perp}}}$, where $L_{g_{W^{\perp}}}$ is the degenerate space of $g_{W^{\perp}}$ in $W^{\perp}$.

One can easily observe from Proposition 1.1.1–(3) that if (W, g_W) is a nondegenerate subspace of index ν' of an inner product space (V, g) of index ν then $(W^{\perp}, g_{W^{\perp}})$ is a nondegenerate subspace of index $\nu - \nu'$ of (V, g). However, note that if (W, g_W) is a degenerate subspace of (V, g) then $(W^{\perp}, g_{W^{\perp}})$ is also a degenerate subspace of (V, g) by Remark 1.1.2. Next we will determine the type of $(W^{\perp}, g_{W^{\perp}})$ when (W, g_W) is degenerate. For this, we need the following lemma which will also be used in the next section.

A subspace W of an inner product space (V, g) is called *totally degenerate* if $g_W = 0$, that is, (W, g_W) is a degenerate subspace of type $(\mu', 0, 0)$. (Hence dim $W = \mu'$).

Lemma 1.1.1 *Let (V, g) be an inner product space and let W be a totally degenerate subspace of (V, g) of dimension $\mu' \geq 1$. Then there exists an orthonormal set of nonnull vectors $\{x_1, y_1, \ldots, x_{\mu'}, y_{\mu'}\}$ in (V, g) such that $\{u_1 = x_1 + y_1, \ldots, u_{\mu'} = x_{\mu'} + y_{\mu'}\}$ is a basis for W.*

Proof. Let x_1, $y_1 \in V$ be orthonormal nonnull vectors with $u_1 = x_1 + y_1 \in W$. (Hence $g(x_1, x_1) = -\, g(y_1, y_1)$). Also let $U_1 = \text{span}\{x_1, y_1\}$ and $U_1^{\perp}$ be the orthogonal space to U_1 in (V, g). Now we claim that $W \cap U_1^{\perp}$ is an $(\mu' - 1)$-dimensional subspace of W. For this, note first that since

$$\text{dim } (W + U_1^{\perp}) + \text{dim } (W \cap U_1^{\perp}) = \text{dim } W + \text{dim } U_1^{\perp},$$

we have

$$\text{dim } (W \cap U_1^{\perp}) = \mu' + n - 2 - \text{ dim } (W + U_1^{\perp}).$$

Hence it suffices to show that dim $(W + U_1^{\perp}) = n - 1$. For this, note first that since $u_1 \in W$, dim $(W + U_1^{\perp}) \geq n - 1$. Yet dim $(W + U_1^{\perp})$ cannot be n, since, if $v \in U_1$ is a null vector which is linearly independent with u_1, then for $w \in W$ and $z \in U_1^{\perp}$, $v = w + z$. Thus v is orthogonal to u_1 in contradiction with the fact that two linearly independent null vectors in U_1 cannot be orthogonal. Now let x_2, $y_2 \in W \cap U_1^{\perp}$, and let $U_2 = \text{span}\{x_1, y_1, x_2, y_2\}$. Then as above, $W \cap U_2^{\perp}$ is a $(\mu' - 2)$-dimensional subspace of W contained in $U_2^{\perp}$. Hence, continuing in this way, we obtain orthonormal vectors $x_1, y_1, \ldots, x_{\mu'}, y_{\mu'} \in V$ with $\{u_1 = x_1 + y_1, \ldots, u_{\mu'} = x_{\mu'} + y_{\mu'}\}$ as a basis for W. □

Proposition 1.1.2 *Let (V, g) be an inner product space of index ν and let (W, g_W) be a degenerate subspace of (V, g) of type (μ', ν', η'). Then $(W^\perp, g_{W^\perp})$ is a degenerate subspace of (V, g) of type $(\mu', \nu - \mu' - \nu', n - \nu - \mu' - \eta')$.*

Proof. First note that, since $L_{g_W} = W \cap W^\perp = L_{g_{W^\perp}}$ by Remark 1.1.2, dim $L_{g_{W^\perp}} = \mu'$. Now let U_1 and U_2 be complementary spaces to L_{g_W} and $L_{g_{W^\perp}}$ in W and $W^\perp$, respectively. (Note that (U_1, g_{U_1}) and (U_2, g_{U_2}) are nondegenerate subspaces). Now suppose $(W^\perp, g_{W^\perp})$ is of type (μ', ν'', η''). Then the restriction of g to $U = U_1 \oplus U_2$ is an inner product of index $\nu' + \nu''$ and $L_{g_{W^\perp}} \subseteq U^\perp$. Since the restriction of g to $U^\perp$ is nondegenerate and $L_{g_{W^\perp}}$ is totally degenerate in $U^\perp$, by Lemma 1.1.1, there exist orthonormal vectors $x_1, y_1, \ldots, x_{\mu'}, y_{\mu'} \in U^\perp$ with $\{u_1 = x_1 + y_1, \ldots, u_{\mu'} = x_{\mu'} + y_{\mu'}\}$ as a basis for $L_{g_{W^\perp}}$. In fact, $\{x_1, y_1, \ldots, x_{\mu'}, y_{\mu'}\}$ spans $U^\perp$, since otherwise, there would be a nonzero nonnull vector $z \in U^\perp$ orthogonal to $\{x_1, y_1, \ldots, x_{\mu'}, y_{\mu'}\}$. But then $z \in W \cap W^\perp$, in contradiction to the fact that $W \cap W^\perp$ is totally degenerate. Thus since $V = U^\perp \oplus U_1 \oplus U_2$, it follows that $\mu' + \nu' + \nu'' = \nu$ and $\mu' + \eta' + \eta'' = n - \nu$. Hence $\nu'' = \nu - \mu' - \nu'$ and $\eta'' = n - \nu - \mu' - \eta'$. □

Proposition 1.1.3 *Let (V, g) be an inner product space of index ν and let (W, g_W) be a degenerate subspace of (V, g) of type (μ', ν', η'). Then $L_{g_W}^\perp = W + W^\perp$ and hence, $(L_{g_W}^\perp, g_{L_{g_W}^\perp})$ is a degenerate subspace of (V, g) of type $(\mu', \nu - \mu', n - \nu - \mu')$.*

Proof. Note that, since $W \cap W^\perp = L_{g_W}$, it follows from Lemma 1.1.1, Proposition 1.1.2 and

$$\dim\,(W + W^\perp) + \dim\,(W \cap W^\perp) = \dim W + \dim W^\perp$$

that

$$\begin{aligned}\dim\,(W + W^\perp) &= \mu' + \nu' + \eta' + \mu' + \nu - \mu' - \nu' + n - \nu - \mu' - \eta' - \mu' \\ &= n - \mu' \\ &= \dim L_{g_W}^\perp.\end{aligned}$$

Hence, since $W + W^\perp \subseteq L_{g_W}^\perp$, $L_{g_W}^\perp = W + W^\perp$. Thus $(L_{g_W}^\perp, g_{L_{g_W}^\perp})$ is degenerate of type $(\mu', \nu' + \nu - \mu' - \nu', \eta' - n - \nu - \mu' - \eta')$, that is, of type $(\mu', \nu - \mu', n - \nu - \mu')$. □

We will make the following observation about degenerate hyperspaces in inner product spaces.

Proposition 1.1.4 *Let (V, g) be an inner product space of index ν and let (W, g_W) be a degenerate hyperspace in (V, g). Then $L_{g_W} = W^{\perp}$ and* $\dim L_{g_W} = 1$. *Hence (W, g_W) is of type $(1, \nu - 1, n - \nu - 1)$.*

Proof. By Proposition 1.1.1–(1), $\dim W^{\perp} = 1$ and $L_{g_W} \subseteq W^{\perp}$. Hence $L_{g_W} = W^{\perp}$. The type of (W, g_W) then follows from Proposition 1.1.3 since $L_{g_W}^{\perp} = W$. □

Finally in this section, we will obtain inner product spaces out of degenerate subspaces by quotiening out the degenerate spaces of these subspaces.

Let (V, g) be an inner product space and let W be a subspace of V. Suppose g_W be of type (μ', ν', η') and let L_{g_W} be the degenerate space of g_W in W. Let $\overline{W} = W/L_{g_W}$ be the quotient space and $\pi_W : W \to \overline{W}$ be the natural projection. We define the *quotient inner product* $\overline{g}_{\overline{W}}$ on $\overline{W}$ by

$$\overline{g}_{\overline{W}}(\overline{x}, \overline{y}) = g_W(x, y),$$

where x, $y \in W$ with $\pi_W(x) = \overline{x}$, $\pi_W(y) = \overline{y}$. To show that $\overline{g}_{\overline{W}}$ is well-defined, first we have to show that $\overline{g}_{\overline{W}}$ is independent of the choice of representatives of $\overline{x}$ and $\overline{y}$. For this, let x, x', y, $y' \in W$ with $\pi_W(x) = \pi_W(x') = \overline{x}$ and $\pi_W(y) = \pi_W(y') = \overline{y}$. Then there exist u, $w \in L_{g_W}$ such that $x = x' + u$ and $y = y' + w$. Thus

$$\overline{g}_{\overline{W}}(\overline{x}, \overline{y}) = g_W(x, y) = g_W(x' + u, y' + w) = g_W(x', y').$$

Note also that $\overline{g}_{\overline{W}}$ is an inner product on $\overline{W}$, that is $L_{\overline{g}_{\overline{W}}} = \{0\}$. To see this, let $\overline{x} \in L_{\overline{g}_{\overline{W}}}$, that is, $\overline{g}_{\overline{W}}(\overline{x}, \overline{y}) = 0$ for every $\overline{y} \in \overline{W}$. Then $g_W(x, y) = 0$ for every $y \in W$, where $x \in W$ with $\pi_W(x) = \overline{x}$. Hence $x \in L_{g_W}$ and it follows that $\overline{x} = \pi_W(x) = 0$. Also observe that $\overline{g}_{\overline{W}}$ is of index ν'.

Let (V, g) be an inner product space of index ν and let W be a subspace of (V, g). A complementary space W' to L_{g_W} in W is called a *geometric realization of* $\overline{W}$. Indeed it can easily be seen that $\pi_W|_{W'}: W' \to \overline{W}$ is a linear isometry between the inner product spaces $(W', g_{W'})$ and $(\overline{W}, \overline{g}_{\overline{W}})$. Note that, if (W, g_W) is nondegenerate then (W, g_W) and $(\overline{W}, \overline{g}_{\overline{W}})$ are canonically isometric by π_W.

Remark 1.1.3 Let (V, g) be an inner product space of index ν and W be a subspace of (V, g). Note that, by Propositions 1.1.2 and 1.1.3, if g_W is of type (μ', ν', η') then $(\overline{W}, \overline{g}_{\overline{W}})$, $(\overline{W^{\perp}}, \overline{g}_{\overline{W^{\perp}}})$ and $(\overline{L_{g_W}^{\perp}}, \overline{g}_{\overline{L_{g_W}^{\perp}}})$ are inner product spaces of indices ν', $\nu - \mu' - \nu'$ and $\nu - \mu'$, respectively. Also note that $\overline{W}$ and $\overline{W^{\perp}}$ are subspaces of $\overline{L_{g_W}^{\perp}}$ and, $\overline{g}_{\overline{W}}$ and $\overline{g}_{\overline{W^{\perp}}}$ are the restrictions of $\overline{g}_{\overline{L_{g_W}^{\perp}}}$ to $\overline{W}$ and $\overline{W^{\perp}}$, respectively. Thus by Proposition 1.1.3, we have the orthogonal direct sum decomposition $\overline{L_{g_W}^{\perp}} = \overline{W} \oplus \overline{W^{\perp}}$. Also note that $\overline{W^{\perp}} = \overline{W}^{\perp}$, where $\overline{W}^{\perp}$ is the orthogonal space to $\overline{W}$ in $(\overline{L_{g_W}^{\perp}}, \overline{g}_{\overline{L_{g_W}^{\perp}}})$.

1.2 Linear Transformations Between Inner Product Spaces

Throughout this section, let (V_1, g_1) and (V_2, g_2) be inner product spaces of dimensions n_1 and n_2, and indices ν_1 and ν_2, respectively.

Let $T : (V_1, g_1) \to (V_2, g_2)$ be a linear transformation. Set

$$\begin{aligned}
\mathcal{V} &= \ker T \\
\mathcal{H} &= (\ker T)^\perp = \mathcal{V}^\perp \\
L_1 &= (\ker T) \cap (\ker T)^\perp = L_{g_{1\mathcal{V}}} = L_{g_{1\mathcal{H}}} \\
A_1 &= \ker T + (\ker T)^\perp = L_1^\perp \\
A_2 &= \text{range } T \\
L_2 &= (\text{range } T) \cap (\text{range } T)^\perp = L_{g_{2A_2}}.
\end{aligned}$$

Let

$$\overline{\mathcal{H}} = \mathcal{H}/L_1 \quad \text{and} \quad \overline{A_2} = A_2/L_2$$

with natural projections

$$\pi_{\mathcal{H}} : \mathcal{H} \to \overline{\mathcal{H}} \quad \text{and} \quad \pi_{A_2} : A_2 \to \overline{A_2}.$$

Also let $\overline{g_{1\mathcal{H}}}$ and $\overline{g_{2A_2}}$ be the quotient inner products on $\overline{\mathcal{H}}$ and $\overline{A_2}$, respectively. Now define a linear transformation

$$\overline{T} : (\overline{\mathcal{H}}, \overline{g_{1\mathcal{H}}}) \to (\overline{A_2}, \overline{g_{2A_2}}),$$

which we call the *quotient of* T, by

$$\overline{T}\overline{x} = \pi_{A_2}(Tx),$$

where $x \in \mathcal{H}$ with $\pi_{\mathcal{H}}(x) = \overline{x}$. To show that $\overline{T}$ is well-defined, it suffices to show that $\overline{T}$ is independent of the choice of a representative of $\overline{x}$ in $\mathcal{H}$. To show this, let $x, x' \in \mathcal{H}$ with $\pi_{\mathcal{H}}(x) = \pi_{\mathcal{H}}(x') = \overline{x}$. Then there exists $u \in L_1$ such that $x = x' + u$. Thus

$$\overline{T}\overline{x} = \pi_{A_2}(Tx) = \pi_{A_2}(T(x' + u)) = \pi_{A_2}(Tx').$$

Remark 1.2.1 By considering the degenerate spaces of $g_{1\mathcal{H}}$ and g_{2A_2}, it can be easily observed that $\overline{T}$ may not be either injective or surjective. Also observe that rank $\overline{T} \leq$ rank T $-$ dim L_1.

Definition 1.2.1 Let $T : (V_1, g_1) \to (V_2, g_2)$ be a linear transformation. The *non-degenerate rank* of T is defined to be the rank of $\overline{T}$.

Next we define the adjoint of a linear transformation between inner product spaces. By making use of the adjoint linear transformation we will introduce an inner product on the space of linear transformations later in Proposition 1.2.2.

Definition 1.2.2 Let $T : (V_1, g_1) \to (V_2, g_2)$ be a linear transformation. The *adjoint* ${}^*T$ of T is defined to be the unique linear transformation ${}^*T : (V_2, g_2) \to (V_1, g_1)$ such that, for all $x \in V_1$ and $y \in V_2$,

$$g_1(x, {}^*Ty) = g_2(Tx, y).$$

Remark 1.2.2 Here is a relation between the adjoint and the transpose of a linear transformation between inner product spaces. Let ${}^tT : V_2^* \to V_1^*$ be the *transpose* of a linear transformation $T : (V_1, g_1) \to (V_2, g_2)$ defined by $({}^tT\omega)x = \omega(Tx)$ for all $\omega \in V_2^*$ and $x \in V_1$. (Cf. [HK, p. 112]). Also let $g_i^\flat : V_i \to V_i^*$, $i = 1, 2$, be the musical isomorphisms defined by $g_i^\flat(x) = g_i(x, \cdot)$. (Note that, since g_i is nondegenerate, $g_i^\flat$ is injective and hence an isomorphism). Let $g_i^\sharp : V_i^* \to V_i$, $i = 1, 2$, be the inverses of $g_i^\flat$, $i = 1, 2$. Then since

$$\begin{aligned} g_2(Tx, y) &= (g_2^\flat(y))(Tx) \\ &= ({}^tT(g_2^\flat(y)))x \end{aligned}$$

and

$$g_1(x, {}^*Ty) = (g_1^\flat({}^*Ty))x$$

for all $x \in V_1$ and $y \in V_2$, it follows from

$$g_1(x, {}^*Ty) = g_2(Tx, y)$$

that

$$g_1^\flat({}^*Ty) = {}^tT(g_2^\flat(y))$$

for every $y \in V_2$. Hence $g_1^\flat \circ {}^*T = {}^tT \circ g_2^\flat$, that is, ${}^*T = g_1^\sharp \circ {}^tT \circ g_2^\flat$.

Next we make the following observation about the kernels and ranges of T and $^*T$.

Proposition 1.2.1 *Let* $T : (V_1, g_1) \to (V_2, g_2)$ *be a linear transformation. Then*

(1) $(\text{range } T)^{\perp} = \ker {}^*T = A_2^{\perp}$

(2) $(\ker T)^{\perp} = \text{range } {}^*T = \mathcal{H}$

Proof. (1) For $y \in V_2$ and for all $x \in V_1$, $y \in (\text{range } T)^{\perp}$ if and only if $g_2(y, Tx) = 0$ if and only if $g_1({}^*Ty, x) = 0$ if and only if ${}^*Ty = 0$ if and only if $y \in \ker {}^*T$.

(2) By applying (1) to ${}^*T : (V_2, g_2) \to (V_1, g_1)$, we have $(\text{range } {}^*T)^{\perp} = \ker {}^*({}^*T) = \ker T$. Hence $(\ker T)^{\perp} = \text{range } {}^*T = \mathcal{H}$. □

Now let $\mathcal{L}(V_1, V_2)$ be the space of linear transformations between inner product spaces (V_1, g_1) and (V_2, g_2). Define a bilinear form $\langle , \rangle$ on $\mathcal{L}(V_1, V_2)$ by

$$\langle T, S \rangle = \text{trace } {}^*T \circ S,$$

where T, $S \in \mathcal{L}(V_1, V_2)$. In fact, $\langle , \rangle$ is an inner product on $\mathcal{L}(V_1, V_2)$.

Proposition 1.2.2 *Let* (V_1, g_1) *and* (V_2, g_2) *be inner product spaces. Then the bilinear form* $\langle , \rangle$ *on* $\mathcal{L}(V_1, V_2)$ *defined by*

$$\langle T, S \rangle = \text{trace } {}^*T \circ S$$

is an inner product on $\mathcal{L}(V_1, V_2)$ *of index* $\nu_1(n_2 - \nu_2) + \nu_2(n_1 - \nu_1)$.

Proof. We will find an orthonormal basis for $\mathcal{L}(V_1, V_2)$ with respect to $\langle , \rangle$ which consists of timelike and spacelike vectors only. Hence the index of $\langle , \rangle$ will also follow from the number of timelike vectors in this orthonormal basis. Now let $\{\alpha_1, \ldots, \alpha_{\nu_1}, \alpha_{\nu_1+1}, \ldots, \alpha_{n_1}\}$ be an orthonormal basis for (V_1, g_1) with

$$g_1(\alpha_i, \alpha_i) = \begin{cases} -1 & \text{for} \quad 1 \le i \le \nu_1 \\ 1 & \text{for} \quad \nu_1 + 1 \le i \le n_1 \end{cases}$$

and let $\{\beta_1, \ldots, \beta_{\nu_2}, \beta_{\nu_2+1}, \ldots, \beta_{n_2}\}$ be an orthonormal basis for (V_2, g_2) with

$$g_2(\beta_i, \beta_i) = \begin{cases} -1 & \text{for} \quad 1 \le i \le \nu_2 \\ 1 & \text{for} \quad \nu_2 + 1 \le i \le n_2 \end{cases}.$$

Now define a basis $\{E_{ij}\}$ for $\mathcal{L}(V_1, V_2)$ by

$$E_{ij}(\alpha_k) = \delta_{ik}\beta_j, \quad 1 \le i \le n_1, \qquad 1 \le j \le n_2,$$

where $k = 1, 2, \ldots, n_1$ and δ_{ik} is the Kronecker delta.

We will show that $\{E_{ij}\}$ is an orthonormal basis for $\mathcal{L}(V_1, V_2)$ with respect to $\langle , \rangle$. Indeed,

$$\begin{aligned}
\langle E_{ij}, E_{lm} \rangle &= \text{trace}\, {}^*E_{ij} \circ E_{lm} \\
&= \sum_{k=1}^{n_1} g_1(\alpha_k, \alpha_k) g_1(({}^*E_{ij} \circ E_{lm})\alpha_k, \alpha_k) \\
&= \sum_{k=1}^{n_1} g_1(\alpha_k, \alpha_k) g_2(E_{lm}\alpha_k, E_{ij}\alpha_k) \\
&= \sum_{k=1}^{n_1} g_1(\alpha_k, \alpha_k) g_2(\delta_{lk}\beta_m, \delta_{ik}\beta_j) \\
&= \sum_{k=1}^{n_1} \delta_{lk}\delta_{ik} g_1(\alpha_k, \alpha_k) g_2(\beta_m, \beta_j) \\
&= \begin{cases} 0 & \text{if} \quad i \ne l \text{ or } j \ne m \\ g_1(\alpha_i, \alpha_i) g_2(\beta_j, \beta_j) & \text{if} \quad i = l \text{ and } j = m. \end{cases}
\end{aligned}$$

Thus $\{E_{ij}\}$ is an orthonormal basis for $\mathcal{L}(V_1, V_2)$ which consists of timelike and spacelike vectors only, that is, $\langle , \rangle$ is an inner product on $\mathcal{L}(V_1, V_2)$. To determine the index of $\langle , \rangle$, it suffices to count the timelike vectors in $\{E_{ij}\}$. For, note that it follows from above that

$$\langle E_{ij}, E_{ij} \rangle = \begin{cases} -1 \quad \text{for} \begin{cases} \{1 \le i \le \nu_1 \text{ and } \nu_2 + 1 \le j \le n_2\} \\ \text{or} \\ \{\nu_1 + 1 \le i \le n_1 \text{ and } 1 \le j \le \nu_2\} \end{cases} \\ 1 \quad \text{for} \begin{cases} \{1 \le i \le \nu_1 \text{ and } 1 \le j \le \nu_2\} \\ \text{or} \\ \{\nu_1 + 1 \le i \le n_1 \text{ and } \nu_2 + 1 \le j \le n_2\} \end{cases} \end{cases}$$

Hence the index of $\langle , \rangle$ is $\nu_1(n_2 - \nu_2) + \nu_2(n_1 - \nu_1)$. □

Definition 1.2.3 Let (V_1, g_1) and (V_2, g_2) be inner product spaces and let $T : (V_1, g_1) \to (V_2, g_2)$ be a linear transformation. Then the *square norm* of T is defined by

$$\|T\|^2 = \langle T, T \rangle.$$

Let $T : (V_1, g_1) \to (V_2, g_2)$ be a linear transformation. Then $\overline{T} : (\overline{\mathcal{H}}, \overline{g_1}_{\overline{\mathcal{H}}}) \to (\overline{A_2}, \overline{g_2}_{\overline{A_2}})$ is also a linear transformation between inner product spaces $(\overline{\mathcal{H}}, \overline{g_1}_{\overline{\mathcal{H}}})$ and $(\overline{A_2}, \overline{g_2}_{\overline{A_2}})$. Let $\|\overline{T}\|^2$ also denote the square norm of $\overline{T}$. We have the following relation between $\|T\|^2$ and $\|\overline{T}\|^2$.

Theorem 1.2.1 *Let $T : (V_1, g_1) \to (V_2, g_2)$ be a linear transformation. Then $\|T\|^2 = \|\overline{T}\|^2$.*

Proof. Let $\{x_1, \ldots, x_l\}$ be an orthonormal basis for a complementary space to L_1 in $\mathcal{V}$ and $\{y_1, \ldots, y_m\}$ be an orthonormal basis for a complementary space to L_1 in $\mathcal{H}$. Also let $\{z_1, w_1, \ldots, z_k, w_k\}$ be an orthonormal basis for (span $\{x_1, \ldots, x_l, y_1, \ldots, y_m\})^{\perp}$ such that $u_i = z_i + w_i \in L_1$ for $i = 1, \ldots, k$. (See the proof of Proposition 1.1.2). Then $\{x_1, \ldots, x_l, y_1, \ldots, y_m, z_1, w_1, \ldots, z_k, w_k\}$ is an orthonormal basis for (V_1, g_1) and we have

$$\begin{aligned}
\|T\|^2 &= \sum_{i=1}^{l} g_1(x_i, x_i) g_1((^*T \circ T)x_i, x_i) + \sum_{i=1}^{m} g_1(y_i, y_i) g_1((^*T \circ T)y_i, y_i) \\
&\quad + \sum_{i=1}^{k} g_1(z_i, z_i) g_1((^*T \circ T)z_i, z_i) + \sum_{i=1}^{k} g_1(w_i, w_i) g_1((^*T \circ T)w_i, w_i) \\
&= \sum_{i=1}^{m} g_1(y_i, y_i) g_2(Ty_i, Ty_i) \\
&\quad + \sum_{i=1}^{k} g_1(z_i, z_i) g_2(Tz_i, Tz_i) + \sum_{i=1}^{k} g_1(w_i, w_i) g_2(Tw_i, Tw_i).
\end{aligned}$$

But since $u_i = z_i + w_i \in L_1 \subseteq \mathcal{V}$,

$$0 = Tu_i = Tz_i + Tw_i$$

and

$$0 = g_1(u_i, u_i) = g_1(z_i, z_i) + g_1(w_i, w_i).$$

Thus $Tz_i = -Tw_i$ and $g_1(z_i, z_i) = -\ g_1(w_i, w_i)$. Hence

$$\begin{aligned}
\|T\|^2 &= \sum_{i=1}^{m} g_1(y_i, y_i) g_2(Ty_i, Ty_i) \\
&= \sum_{i=1}^{m} \overline{g_1}_{\overline{\mathcal{H}}}(\overline{y}_i, \overline{y}_i) \overline{g_2}_{\overline{A_2}}(\overline{T}\overline{y}_i, \overline{T}\overline{y}_i) \\
&= \|\overline{T}\|^2, \text{ where } \pi_{\mathcal{H}}(y_i) = \overline{y}_i.
\end{aligned}$$

□

Now we define certain linear transformations between inner product spaces which relate the square norm and the nondegenerate rank. These linear transformations may also be considered as partial isometries. In later chapters, we will generalize these transformations to maps between semi-Riemannian manifolds.

Definition 1.2.4 Let $T : (V_1, g_1) \to (V_2, g_2)$ be a linear transformation. Then T is called *semi-Euclidean* if $\overline{T} : (\overline{\mathcal{H}}, \overline{g}_{\overline{\mathcal{H}}}) \to (\overline{A_2}, \overline{g_2}_{\overline{A_2}})$ is an (into) isometry.

Remark 1.2.3 Let $T : (V_1, g_1) \to (V_2, g_2)$ be a semi-Euclidean linear transformation. Then since $\overline{T}$ is injective (yet may not be surjective), rank $\overline{T}$ = rank T − dim L_1 and hence, dim $\overline{\mathcal{H}}$ = rank $\overline{T}$.

Theorem 1.2.2 *Let $T : (V_1, g_1) \to (V_2, g_2)$ be a semi-Euclidean linear transformation. Then T satisfies the equation*

$$\|T\|^2 = \text{rank } \overline{T}.$$

Proof. Let ${}^*\overline{T}$ be the adjoint of $\overline{T}$ and define a linear transformation

$$\overline{F} : (\overline{\mathcal{H}}, \overline{g}_{\overline{\mathcal{H}}}) \to (\overline{\mathcal{H}}, \overline{g}_{\overline{\mathcal{H}}})$$

by

$$\overline{F} = {}^*\overline{T} \circ \overline{T}.$$

First we will show that $\overline{F}$ is the identity linear transformation. Indeed, note that for every $\overline{x}$, $\overline{y} \in \overline{\mathcal{H}}$,

$$\overline{g_1}_{\overline{\mathcal{H}}}(\overline{F}\overline{x}, \overline{y}) = \overline{g_2}_{\overline{A_2}}(\overline{T}\overline{x}, \overline{T}\overline{y}) = \overline{g}_{\overline{\mathcal{H}}}(\overline{x}, \overline{y}).$$

Hence $\overline{F} = \overline{Id}$ and it follows from Theorem 1.2.1 and Remark 1.2.3 that

$$\|T\|^2 = \|\overline{T}\|^2 = \text{trace}\, \overline{F} = \dim \overline{\mathcal{H}} = \text{rank}\, \overline{T}.$$

□

Now it is worth giving a typical example of a semi-Euclidean linear transformation. Let

$$V_1 = \mathbf{R}^4, \quad g_1((x^1, x^2, x^3, x^4), (y^1, y^2, y^3, y^4)) = -x^1y^1 - x^2y^2 + x^3y^3 + x^4y^4$$

and let

$$V_2 = \mathbf{R}^3, \quad g_2((z^1, z^2, z^3), (w^1, w^2, w^3)) = -z^1w^1 - z^2w^2 + z^3w^3.$$

Then it can be shown that the linear transformation $T : (V_1, g_1) \to (V_2, g_2)$ which is given by

$$T(x^1, x^2, x^3, x^4) = (x^1 + x^3, x^1 + x^3, x^4)$$

is semi-Euclidean with dim $L_1 = 1$, rank $T = 2$ and rank $\overline{T} = 1$. Since this linear transformation is semi-Euclidean, it satisfies $\|T\|^2 = \text{rank}\, \overline{T} = 1$.

Definition 1.2.5 *Let $T : (V_1, g_1) \to (V_2, g_2)$ be a linear transformation. Then T is called weakly semi-Euclidean if $\|T\|^2 = \text{rank}\, \overline{T}$.*

In general, a weakly semi-Euclidean linear transformation is not necessarily semi-Euclidean. For example, let

$$V_1 = \mathbf{R}^3, \quad g_1((x^1, x^2, x^3), (y^1, y^2, y^3)) = -x^1y^1 - x^2y^2 + x^3y^3$$

and let

$$V_2 = \mathbf{R}^2, \quad g_2((z^1, z^2), (w^1, w^2)) = z^1w^1 + z^2w^2.$$

Then it can be shown that the linear transformation $T : (V_1, g_1) \to (V_2, g_2)$ which is given by

$$T(x^1, x^2, x^3) = (x^1 + \sqrt{2}x^3, x^2 + \sqrt{2}x^3)$$

satisfies $\|T\|^2 = \text{rank}\, \overline{T} = \text{rank}\, T = 2$. Yet note that T is not a semi-Euclidean linear transformation since $(\overline{\mathcal{H}}, \overline{g_{1\overline{\mathcal{H}}}})$ cannot be isometric to $(\overline{A_2}, \overline{g_{2\overline{A_2}}})$ because of their different indices. However we have the following partial converse to Theorem 1.2.2.

Theorem 1.2.3 *Let $T : (V_1, g_1) \to (V_2, g_2)$ be a linear transformation, where $n_2 =$ dim $V_2 = 1$. If T is weakly semi-Euclidean then T is semi-Euclidean.*

Proof. Note that $0 \leq \text{rank } \overline{T} \leq 1$. Now first suppose that rank $\overline{T} = 0$. If $\mathcal{V}$ = ker $T = V_1$ then, since $\mathcal{H} = \{0\}$, $T = 0$. Thus $\overline{T} : (\overline{\mathcal{H}} = \{0\}, \overline{g_1}_{\overline{\mathcal{H}}} = \{0\}) \to (\overline{A_2} = \{0\}, \overline{g_2}_{\overline{A_2}} = \{0\})$ is an isometry. If $\mathcal{V}$ = ker T is a degenerate hyperspace in (V_1, g_1) then by Proposition 1.1.4, $\mathcal{H} = L_1$, and hence $\overline{\mathcal{H}} = \{0\}$. Thus $\overline{T} : (\overline{\mathcal{H}} = \{0\}, \overline{g_1}_{\overline{\mathcal{H}}} = \{0\}) \to (\overline{A_2}, \overline{g_2}_{\overline{A_2}})$ is an into isometry. Note that $\mathcal{V}$ = ker T cannot be a nondegenerate hyperspace in (V_1, g_1) for the case rank $\overline{T} = 0$. Now suppose that rank $\overline{T} = 1$. Then by Remark 1.2.1, dim $L_1 = 0$ and hence $\mathcal{V}$ = ker T is a nondegenerate hyperspace in (V_1, g_1). Thus $(\mathcal{H}, g_{1\mathcal{H}})$ is a 1-dimensional inner product space and if $x \in \mathcal{H}$ is a unit vector,

$$1 = \|T\|^2 = g_1(x, x) g_2(Tx, Tx) = \overline{g_1}(\overline{x}, \overline{x}) \overline{g_2}(\overline{T}\overline{x}, \overline{T}\overline{x}),$$

where $\pi_{\mathcal{H}}(x) = \overline{x}$. Thus

$$\overline{g_2}(\overline{T}\overline{x}, \overline{T}\overline{x}) = \overline{g_1}(\overline{x}, \overline{x}),$$

that is, $\overline{T} : (\overline{\mathcal{H}}, \overline{g}_{\overline{\mathcal{H}}}) \to (\overline{A_2}, \overline{g_2}_{\overline{A_2}})$ is an isometry. □

Let (V, g) be an inner product space of dimension n and index ν. A linear transformation $T : (V, g) \to (V, g)$ is called *self-adjoint* if $T = {}^*T$. If $0 < \nu < n$ then in general, a self-adjoint linear transformation T on (V, g) may not be diagonalizable. But it can easily be shown that if T is a diagonalizable self-adjoint linear transformation on (V, g) then the eigenspaces of T are nondegenerate and orthogonal to each other.

Lemma 1.2.1 *Let $T : (V, g) \to (V, g)$ be a diagonalizable self-adjoint linear transformation. Then*

$$\|T\|^2 \geq \frac{1}{n}(\text{trace}\, T)^2$$

Proof. Let $\{e_i\}$ be an orthonormal basis for (V, g) consists of eigenvectors of T with corresponding eigenvalues $\{\lambda_i\}$. Then

$$\begin{aligned}
\|T\|^2 &= \sum_{i=1}^{n} g(e_i, e_i) g(Te_i, Te_i) \\
&= \sum_{i=1}^{n} \lambda_i^2 g(e_i, e_i)^2 \\
&= \sum_{i=1}^{n} \lambda_i^2 \\
&\geq \frac{1}{n}(\sum_{i=1}^{n} \lambda_i)^2 \\
&= \frac{1}{n}(\sum_{i=1}^{n} g(e_i, e_i) g(Te_i, e_i))^2 \\
&= \frac{1}{n}(\mathrm{trace}\, T)^2.
\end{aligned}$$

□

Finally we will make some further observations on semi-Euclidean linear transformations between positive definite inner product spaces.

Let $T : (V_1, g_1) \to (V_2, g_2)$ be a linear transformation between positive definite inner product spaces (V_1, g_1) and (V_2, g_2). Then note that, since $(\mathcal{H}, g_{1\mathcal{H}})$ and (A_2, g_{2A_2}) are positive definite inner product spaces, $(\mathcal{H}, g_{1\mathcal{H}})$ and (A_2, g_{2A_2}) are canonically isometric to $(\overline{\mathcal{H}}, \overline{g_{1\overline{\mathcal{H}}}})$ and $(\overline{A_2}, \overline{g_{2\overline{A_2}}})$ by the natural projections $\pi_{\mathcal{H}}$ and π_{A_2}, respectively. Hence by identifying $(\mathcal{H}, g_{1\mathcal{H}})$ and (A_2, g_{2A_2}) with $(\overline{\mathcal{H}}, \overline{g_{1\overline{\mathcal{H}}}})$ and $(\overline{A_2}, \overline{g_{2\overline{A_2}}})$, respectively, we can identify the quotient $\overline{T} : \overline{\mathcal{H}} \to \overline{A_2}$ of T by

$$T^h : \mathcal{H} \to A_2$$

defined by

$$T^h x = Tx,$$

where $x \in \mathcal{H}$. (Here, it is important to observe that T^h is an isomorphism and hence the nondegenerate rank of T is equal to the rank of T^h which is equal to the rank of T). Considering T^h as a linear transformation $T^h : (\mathcal{H}, g_{1\mathcal{H}}) \to (A_2, g_{2A_2})$, we denote the adjoint of T^h by $^*(T^h)$. Also by Proposition 1.2.1, the linear transformation

$$(^*T)^h : A_2 \to \mathcal{H}$$

defined by

$$({}^*T)^h y = {}^*Ty,$$

where $y \in A_2$, is an isomorphism. In fact, ${}^*(T^h)$ and $({}^*T)^h$ are the same linear transformations.

Proposition 1.2.3 [F1] *Let $T : (V_1, g_1) \to (V_2, g_2)$ be a linear transformation between positive definite inner product spaces (V_1, g_1) and (V_2, g_2). Then ${}^*(T^h) = ({}^*T)^h$.*

Proof. Let $y \in A_2$. Then for every $x \in \mathcal{H}$,

$$g_1(x, ({}^*T)^h y) = g_1(x, {}^*Ty) = g_1(Tx, y) = g_1(T^h x, y) = g_1(x, {}^*(T^h) y).$$

Thus for every $y \in A_2$, ${}^*(T^h)y = ({}^*T)^h y$, that is ${}^*(T^h) = ({}^*T)^h$. □

Now let $T : (V_1, g_1) \to (V_2, g_2)$ be a linear transformation between inner product spaces (V_1, g_1) and (V_2, g_2) of indices ν_1 and ν_2, respectively. Define linear transformations

$$P : (V_1, g_1) \to (V_1, g_1) \quad \text{by} \quad P = {}^*T \circ T$$

and

$$Q : (V_2, g_2) \to (V_2, g_2) \quad \text{by} \quad Q = T \circ {}^*T.$$

Clearly, by definition P and Q are self-adjoint linear transformations on (V_1, g_1) and (V_2, g_2), respectively. Also it can easily be seen from Proposition 1.2.1 that

$$\begin{aligned} \ker P &= \mathcal{V} = \mathcal{H}^\perp \\ \operatorname{range} P &= \mathcal{H} \end{aligned}$$

and

$$\begin{aligned} \ker Q &= A_2^\perp \\ \operatorname{range} Q &= A_2. \end{aligned}$$

In particular, if $\nu_1 = 0 = \nu_2$, that is, (V_1, g_1) and (V_2, g_2) are positive definite inner product spaces, define the isomorphisms

$$P^h : \mathcal{H} \to \mathcal{H} \quad \text{by} \quad P^h x = Px$$

and

$$Q^h : A_2 \to A_2 \quad \text{by} \quad Q^h y = Qy,$$

where $x \in \mathcal{H}$ and $y \in A_2$. Note that in this case, from Proposition 1.2.3,

$$P^h = ({}^*T \circ T)^h = ({}^*T)^h \circ T^h = {}^*(T^h) \circ T^h$$

and

$$Q^h = (T \circ {}^*T)^h = T^h \circ ({}^*T)^h = T^h \circ {}^*(T^h).$$

Definition 1.2.6 A semi-Euclidean linear transformation $T : (V_1, g_1) \to (V_2, g_2)$ between positive definite inner product spaces (V_1, g_1) and (V_2, g_2) is called a *Euclidean linear transformation.* In other words, a linear transformation $T : (V_1, g_1) \to (V_2, g_2)$ between positive definite inner product spaces (V_1, g_1) and (V_2, g_2) is called *Euclidean* if $T^h : (\mathcal{H}, g_{1\mathcal{H}}) \to (A_2, g_{2A_2})$ is an isometry.

Remark 1.2.4 Note that if $T : (V_1, g_1) \to (V_2, g_2)$ is a Euclidean linear transformation then $\|T\|^2 = \text{rank}\, T$ by Theorem 1.2.2 since $\text{rank}\, \overline{T} = \text{rank}\, T$.

Now we give a characterization of Euclidean linear transformations.

Proposition 1.2.4 [F1] *Let $T : (V_1, g_1) \to (V_2, g_2)$ be a linear transformation between positive definite inner product spaces (V_1, g_1) and (V_2, g_2). Then the following are equivalent:*

(1) *T is Euclidean.*

(2) *P is a projection, that is, $P \circ P = P$.*

(3) *Q is a projection, that is, $Q \circ Q = Q$.*

(4) *$P^h = id_{\mathcal{H}}$, that is, P^h is the identity on $\mathcal{H}$.*

(5) *$Q^h = id_{A_2}$, that is, Q^h is the identity on A_2.*

(6) $(T^h)^{-1} = ({}^*T)^h = {}^*(T^h)$.

Proof. (1) $\Leftrightarrow$ (3): T is Euclidean if and only if for every x_1, $x_2 \in \mathcal{H}$,

$$g_1(x_1, x_2) = g_2(T^h x_1, T^h x_2) = g_2(Tx_1, Tx_2).$$

Since $\mathcal{H} = \text{range } {}^*T$ by Proposition 1.2.1–(2), this equation holds for every $x_1, x_2 \in \mathcal{H}$ if and only if for every $y_1, y_2 \in V_2$,

$$g_1({}^*Ty_1, {}^*Ty_2) = g_2((T \circ {}^*T)y_1, (T \circ {}^*T)y_2),$$

which is equivalent to

$$g_2(y_1, (T \circ {}^*T)y_2) = g_2(y_1, (T \circ {}^*T \circ T \circ {}^*T)y_2)$$

for every $y_1, y_2 \in V_2$. Since $Q = T \circ {}^*T$, this equation is equivalent to

$$g_2(y_1, Qy_2) = g_2(y_1, (Q \circ Q)y_2)$$

for every $y_1, y_2 \in V_2$. Thus $Q = Q \circ Q$.

(3) $\Leftrightarrow$ (5): A linear transformation A on a real vector space V is a projection if and only if the restriction of A on (range A) is the identity map. Since Q is self-adjoint, range $Q = (\ker Q)^\perp$, so the restriction of Q to (range Q) $= A_2$ is precisely the linear transformation Q^h. Since (range Q) = (range T) $= A_2$, Q is a projection if and only if $Q^h = \mathrm{id}_{A_2}$.

(1) $\Leftrightarrow$ (4): From the first equation in ((1) $\Leftrightarrow$ (3)), T^h is an isometry if and only if for every $x_1, x_2 \in \mathcal{H}$,

$$\begin{aligned} g_1(x_1, x_2) &= g_2(Tx_1, Tx_2) = g_1(x_1, ({}^*T \circ T)x_2) \\ &= g_1(x_1, Px_2) = g_1(x_1, P^h x_2). \end{aligned}$$

Hence, since (range P^h) $= \mathcal{H}$, this is equivalent to $P^h = \mathrm{id}_{\mathcal{H}}$.

(2) $\Leftrightarrow$ (4): As for the case ((3) $\Leftrightarrow$ (5)) above, the restriction of P to (range P) $= \mathcal{H}$ is the linear transformation P^h (since (range P) = (range ${}^*T$) $= \mathcal{H}$), and so P is a projection if and only if $P^h = \mathrm{id}_{\mathcal{H}}$.

(4) $\Leftrightarrow$ (6): $P^h = ({}^*T)^h \circ T^h = \mathrm{id}_{\mathcal{H}}$ if and only if $(T^h)^{-1} = ({}^*T)^h = {}^*(T^h)$, where the second equality is given by Proposition 1.2.3. □

Chapter 2

Semi-Riemannian Manifolds

In this chapter, we introduce the concept of a semi-Riemannian manifold and develop the necessary machinery to analyze its geometric structure. In section 2.1, we define the metric tensor on a manifold. Then in section 2.2, we introduce connections on a semi-Riemannian manifold compatible with the metric tensor. We also define parallel translation and geodesics determined by the semi-Riemannian structure of a manifold. In section 2.3 we study the curvatures of the Levi-Civita connection of a semi-Riemannian manifold. In section 2.4, we construct some differential operators on a semi-Riemannian manifold such as, divergence, gradient, Hessian tensor and Laplacian. Finally in section 2.5 we state the semi-Riemannian divergence theorem.

In this chapter, we assume that every object at hand is smooth, unless otherwise stated. Also, we assume that the manifolds are second countable.

2.1 Metric Tensors

Throughout this section, let M be an n-dimensional manifold. Let T_pM denote the tangent space of M at $p \in M$ and $TM = \bigcup_{p\in M} T_pM$ be the tangent bundle of M. As usual, let T_pM^* be the dual space to T_pM and $TM^* = \bigcup_{p\in M} T_pM^*$ be the cotangent bundle of M. We denote the set of sections of TM by ΓTM, that is, the set of vector fields on M. Also we denote the sections of TM^* by ΓTM^*, that is, the set of 1-forms on M. It is important to observe that ΓTM and ΓTM^* are modules over the ring of smooth functions $C^\infty(M)$ on M.

Definition 2.1.1 Let M be a manifold. A multilinear map

$$T : (\times_{i=1}^{r} \Gamma TM^*) \times (\times_{i=1}^{s} \Gamma TM) \to C^\infty(M)$$

is called an (r, s)-*tensor field* on M.

Note that the set $T^r_s(M)$ of tensor fields of type (r,s) on M is also a module over $C^\infty(M)$. The exceptional case $r = s = 0$, a tensor field of type $(0,0)$ on M is just a function $h \in C^\infty(M)$, that is, $T^0_0(M) = C^\infty(M)$. Also, an alternating $(0,s)$-tensor field on M is called an *exterior s-form* on M. The set of exterior s-forms on M will be denoted by $\wedge^s(M)$.

As a terminology, a set of vector fields $\{Y_1, \dots, Y_n\}$ defined on an open subset U of a manifold M is called a *local frame* (for TM) if $\{Y_1(p), \dots, Y_n(p)\}$ is a basis for T_pM at each $p \in U$. Similarly, a set of 1-forms $\{\omega_1, \dots, \omega_n\}$ defined on an open subset U of a manifold M is called a *local co-frame* (a local frame for TM^*) if $\{\omega_1(p), \dots, \omega_n(p)\}$ is a basis for T_pM^* at each $p \in U$.

Proposition 2.1.1 *Let M be a manifold and $T \in T^r_s(M)$. Let $\theta'_1, \dots, \theta'_r$ and $\theta_1, \dots, \theta_r$ be 1-forms on M such that $\theta'_i(p) = \theta_i(p)$, $i = 1, \dots, r$, and let $X'_1, \dots, X'_s$ and $X_1, \dots, X_s$ be vector fields on M such that $X'_i(p) = X_i(p)$, $i = 1, \dots, s$. Then*

$$T(\theta'_1, \dots, \theta'_r, X'_1, \dots, X'_s)(p) = T(\theta_1, \dots, \theta_r, X_1, \dots, X_s)(p)$$

Proof. Let $\{\omega_1, \dots, \omega_n\}$ be a local co-frame and $\{Y_1, \dots, Y_n\}$ be a local frame on a neighborhood U of p. Then

$$\theta'_i = \sum_{j_i=1}^{n} a'^{j_i}_i \omega_{j_i} \quad \text{and} \quad \theta_i = \sum_{j_i=1}^{n} a^{j_i}_i \omega_{j_i} \quad \text{with } a'^{j_i}_i(p) = a^{j_i}_i(p)$$

and

$$X'_i = \sum_{k_i=1}^{n} b'^{k_i}_i Y_{k_i} \quad \text{and} \quad X_i = \sum_{k_i=1}^{n} b^{k_i}_i Y_{k_i} \quad \text{with } \; b'^{k_i}_i(p) = b^{k_i}_i(p),$$

where $a'^{j_i}_i$, $a^{j_i}_i$, $b'^{k_i}_i$, $b^{k_i}_i$ are some functions defined on U. Hence, by linearity

$$T(\theta'_1,\ldots,\theta'_r,X'_1,\ldots,X'_s)(p)$$

$$= \left(\sum_{\substack{j_1,\ldots,j_r=1\\ k_1,\ldots,k_s=1}}^{n} a'^{j_1}_1 \cdots a'^{j_r}_r b'^{k_1}_1 \cdots b'^{k_s}_s T(\omega_{j_1},\ldots,\omega_{j_r},Y_{k_1},\ldots,Y_{k_s}) \right)(p)$$

$$= \sum_{\substack{j_1,\ldots,j_r=1\\ k_1,\ldots,k_s=1}}^{n} a'^{j_1}_1(p) \cdots a'^{j_r}_r(p) b'^{k_1}_1(p) \cdots b'^{k_s}_s(p) T(\omega_{j_1},\ldots,\omega_{j_r},Y_{k_1},\ldots,Y_{k_s})(p)$$

$$= \sum_{\substack{j_1,\ldots,j_r=1\\ k_1,\ldots,k_s=1}}^{n} a^{j_1}_1(p) \cdots a^{j_r}_r(p) b^{k_1}_1(p) \cdots b^{k_s}_s(p) T(\omega_{j_1},\ldots,\omega_{j_r},Y_{k_1},\ldots,Y_{k_s})(p)$$

$$= \left(\sum_{\substack{j_1,\ldots,j_r=1\\ k_1,\ldots,k_s=1}}^{n} a^{j_1}_1 \cdots a^{j_r}_r b^{k_1}_1 \cdots b^{k_s}_s T(\omega_{j_1},\ldots,\omega_{j_r},Y_{k_1},\ldots,Y_{k_s}) \right)(p)$$

$$= T(\theta_1,\ldots,\theta_r,X_1,\ldots,X_s)(p)$$

□

Proposition 2.1.1 allows us to define the value of an (r,s)-tensor field on M at a point $p \in M$.

Definition 2.1.2 Let M be a manifold and $T \in T^r_s(M)$. Then the *value* of T at $p \in M$ is defined to be the multilinear map

$$T_p : (\times^r_{i=1} T_pM^*) \times (\times^s_{i=1} T_pM) \to \mathbf{R}$$

by

$$T_p(\alpha_1,\ldots,\alpha_r,x_1,\ldots,x_s) = T(\theta_1,\ldots,\theta_r,X_1,\ldots,X_s)(p),$$

where $\{\theta_i\}$ are 1-forms on M with $\theta_i(p) = \alpha_i$, $i = 1,\ldots,r$, and $\{X_i\}$ are vector fields on M with $X_i(p) = x_i$, $i = 1,\ldots,s$.

Notice that T_p is well-defined by Proposition 2.1.1. Also multilinearity of T_p follows from the multilinearity of T.

Remark 2.1.1 Note that, by the above definition, we may also consider an (r,s)-tensor field T on a manifold M as a map assigning each point $p \in M$ to a multilinear map $T_p : (\times^r_{i=1} T_pM^*) \times (\times^s_{i=1} T_pM) \to \mathbf{R}$.

Definition 2.1.3 Let M be a manifold. A *metric tensor* g *(of index* ν*)* on M is a $(0,2)$-tensor field on M such that g_p is an inner product of index ν on each T_pM for $p \in M$. If g is a metric tensor (of index ν) on M then (M, g) is called a *semi-Riemannian manifold (of index* ν*)*. In particular, a semi-Riemannian manifold of index $\nu = 0$ (resp., $\nu = 1$) is called a *Riemannian* (resp., *Lorentzian*) *manifold.*

A symmetric $(0,2)$-tensor field g on M is called *nondegenerate of index* ν *at* $p \in M$ if g_p is an inner product on T_pM of index ν.

Lemma 2.1.1 *Let M be a manifold and g be a symmetric $(0,2)$-tensor field on M. Then the set of points where g is nondegenerate of index ν is an open subset of M, where $0 \leq \nu \leq n$.*

Proof. If the set of points where g is nondegenerate of index ν $(0 \leq \nu \leq n)$ is empty, then the claim follows trivially. Suppose the set of points where g is nondegenerate of index ν is not empty for some $0 \leq \nu \leq n$. We show that any point p in this set has a neighborhood such that g is nondegenerate of index ν at every point of this neighborhood. For, let g be nondegenerate of index ν at $p \in M$. Then there exists an orthonormal basis $\{x_1, \ldots, x_\nu, x_{\nu+1}, \ldots, x_n\}$ for (T_pM, g_p) such that

$$g(x_i, x_i) = \begin{cases} -1 & \text{for} \quad 1 \leq i \leq \nu \\ \\ 1 & \text{for} \quad \nu + 1 \leq i \leq n \end{cases}$$

Extend $\{x_1, \ldots, x_\nu, x_{\nu+1}, \ldots, x_n\}$ to a local frame $\{X_1, \ldots, X_\nu, X_{\nu+1}, \ldots, X_n\}$ on a neighborhood U of p and let g_{ij} be functions on U defined by

$$g_{ij} = g(X_i, X_j), \qquad 1 \leq i, j \leq n.$$

First note that, since $\det[g_{ij}(p)] = (-1)^\nu$, $\det[g_{ij}] \neq 0$ on a neighborhood of p in U. Hence by shrinking U to this neighborhood if necessary, we may assume that g_q is nondegenerate of some index ν_q at each point $q \in U$. Also since g_{ij}, $1 \leq i, j \leq n$, are continuous functions on U, for $\varepsilon_+ = \frac{1}{n-\nu+2}$, there exists a neighborhood U_+ of p in U such that

$$| \, g_{ii}(q) - 1 \, | < \varepsilon_+ \quad \text{for } \nu + 1 \leq i \leq n$$

and

$$| \, g_{ij}(q) \, | < \varepsilon_+ \quad \text{for } i \neq j, \, \nu + 1 \leq i, j \leq n$$

whenever $q \in U_+$. Now let $W_{+_q} = \operatorname{span}\{X_{\nu+1}(q), \ldots, X_n(q)\}$ at each $q \in U_+$. We show that $g(x, x) > 0$ for every $0 \neq x \in W_{+_q}$ at each $q \in U_+$. Indeed, for any $0 \neq$
$x = \sum\limits_{i=\nu+1}^{n} \lambda_i X_i(q) \in W_{+_q}$,

$$\begin{aligned}
g(x,x) &= \sum_{i,j=\nu+1}^{n} \lambda_i\lambda_j g_{ij}(q) \\
&= \sum_{i=\nu+1}^{n} \lambda_i^2 g_{ii}(q) + 2 \sum_{\nu+1\le i<j\le n}^{n} \lambda_i\lambda_j g_{ij}(q) \\
&> \sum_{i=\nu+1}^{n} \lambda_i^2(1-\varepsilon_+) - 2 \sum_{\nu+1\le i<j\le n}^{n} |\lambda_i||\lambda_j|\varepsilon_+ \\
&= \sum_{i=\nu+1}^{n} \lambda_i^2 - \varepsilon_+ \sum_{i=\nu+1}^{n} \lambda_i^2 - 2\varepsilon_+ \sum_{\nu+1\le i<j\le n}^{n} |\lambda_i||\lambda_j| \\
&\ge \sum_{i=\nu+1} \lambda_i^2 - \varepsilon_+ \sum_{i=\nu+1}^{n} \lambda_i^2 - \varepsilon_+ \sum_{\nu+1\le i<j\le n}^{n} (\lambda_i^2+\lambda_j^2) \\
&= \sum_{i=\nu+1}^{n} \lambda_i^2 - \varepsilon_+(2\lambda_{\nu+1}^2 + 3\lambda_{\nu+2}^2 + \cdots + (n-\nu+1)\lambda_n^2) > 0.
\end{aligned}$$

Similarly it can be shown that there exists a neighborhood U_- of p in U such that $g(x,x) < 0$ for every $x \in W_{-_q}$ at each $q \in U_-$, where $W_{-_q} = \text{span}\{X_1(q), \ldots, X_\nu(q)\}$. Hence on $U' = U_+ \cap U_-$, g_q is positive definite on W_{+_q} and negative definite on W_{-_q} at each $q \in U'$. Then by Remark 1.1.1, $n - \nu_q \ge n - \nu$ and $\nu_q \ge \nu$ at each $q \in U'$. Thus $\nu_q = \nu$ for every $q \in U'$, that is, g is nondegenerate of index ν at every $q \in U'$. □

Corollary 2.1.1 *Let M be a manifold and g be a symmetric $(0,2)$-tensor field on M. Then the set of points where g_p is nondegenerate is an open subset of M.*

Proof. Immediate from Lemma 2.1.1. □

Theorem 2.1.1 *Let M be a connected manifold. If g is a symmetric $(0,2)$-tensor field on M such that g_p is nondegenerate of index ν_p at each $p \in M$ then $\nu_p = \nu$ (constant) for every $p \in M$, that is, g is a metric tensor of index ν on M.*

Proof. Let g_p be of index ν_p at $p \in M$. Then by Lemma 2.1.1, the set of points $q \in M$ where g_q has index ν_p is open in M. Again by Lemma 2.1.1, the complement of this set is also open. Thus by the connectedness of M, g is of index $\nu_p = \nu$ (constant) at every $p \in M$. □

From the viewpoint of vector bundles, a metric tensor on a manifold is in fact a metric tensor in the tangent bundle (which is a vector bundle). The concepts introduced in this section can be generalized to vector bundles. Now we briefly state some of these generalizations which will be referred to from time to time.

Let E be a real vector bundle over a manifold M and let E_p denote the fiber of E at $p \in M$. Also let E^* be the dual bundle of E defined by $E^* = \bigcup_{p\in M} E_p^*$. Let ΓE and ΓE^* denote the set of sections of E and E^*, respectively. It is important to observe that ΓE and ΓE^* are modules over the ring $C^\infty(M)$.

Definition 2.1.4 *Let E be a real vector bundle over a manifold M. A multilinear map*

$$T : (\times_{i=1}^{r}\Gamma E^*) \times (\times_{i=1}^{s}\Gamma E) \to C^\infty(M)$$

is called an (r,s)-tensor field in E.

Note that the set $T_s^r(E)$ of tensor fields of type (r,s) in E is also a module over $C^\infty(M)$. Also Proposition 2.1.1 remains valid for (r,s)-tensor fields in E, that is, if $\theta_1, \ldots, \theta_r, \theta'_1, \ldots, \theta'_r \in \Gamma E^*$ with $\theta_i(p) = \theta'_i(p)$, $i = 1, \ldots, r$, and $X_1, \ldots, X_s, X'_1, \ldots, X'_s \in \Gamma E$ with $X_i(p) = X'_i(p)$, $i = 1, \ldots, s$, then

$$T(\theta'_1, \ldots, \theta'_r, X'_1, \ldots, X'_s)(p) = T(\theta_1, \ldots, \theta_r, X_1, \ldots, X_s)(p).$$

As in Definition 2.1.2, this allows us to define the *value* of $T \in T_s^r(E)$ at $p \in M$ to be the multilinear map

$$T_p : (\times_{i=1}^{r}E_p^*) \times (\times_{i=1}^{s}E_p) \to \mathbf{R}$$

by

$$T_p(\alpha_1, \ldots, \alpha_r, x_1, \ldots, x_s) = T(\theta_1, \ldots, \theta_r, X_1, \ldots, X_s)(p),$$

where $\theta_1, \ldots, \theta_r \in \Gamma E^*$ with $\theta_i(p) = \alpha_i$, $i = 1, \ldots, r$, and $X_1, \ldots, X_s \in \Gamma E$ with $X_i(p) = x_i$, $i = 1, \ldots, s$.

By using these generalizations, we can define metric tensors in real vector bundles.

Definition 2.1.5 Let E be a real vector bundle over a manifold M. A *metric tensor* g *(of index ν)* in E is a $(0,2)$-tensor field in E such that g_p is an inner product of index ν on each E_p for $p \in M$. If g is a metric tensor (of index ν) in E then (E, g) is called a *semi-Riemannian vector bundle (of index ν)* over M.

A symmetric $(0,2)$-tensor field g in E is called *nondegenerate of index ν at* $p \in M$ if g_p is an inner product on E_p of index ν. It is also easy to observe that Lemma

2.1.1 remains valid for real vector bundles. That is, if E is a vector bundle over a manifold M and g is a symmetric $(0,2)$-tensor field in E, then the set of points where g_p is nondegenerate of index ν is an open subset of M. Consequently, we also have the following theorem as a generalization of Theorem 2.1.1.

Theorem 2.1.2 *Let E be a real vector bundle over a connected manifold M. If g is a symmetric $(0,2)$-tensor field in E such that g_p is nondegenerate of index ν_p at each $p \in M$, then $\nu_p = \nu$ (constant) for every $p \in M$, that is, g is a metric tensor (of index ν) in E.*

Finally, let E' be a subbundle of a semi-Riemannian vector bundle (E, g) over a manifold M. If the restriction $g_{E'}$ of g to E' is a metric tensor in E' (of index ν') then we call $(E', g_{E'})$ a *semi-Riemannian subbundle (of index ν')* of (E, g).

2.2 Connections on Semi-Riemannian Manifolds

Throughout this section, let M, M_1 and M_2 be manifolds of dimensions n, n_1 and n_2, respectively.

Definition 2.2.1 Let $f : M_1 \to M_2$ be a map. Then a map $X : M_1 \to TM_2$ is called a *vector field along* f if $X(p_1) \in T_{f(p_1)}M_2$ for each $p_1 \in M_1$.

We denote the set of vector fields along $f : M_1 \to M_2$ by $\Gamma_f TM_2$. Note that $\Gamma_f TM_2$ is a module over the ring $C^\infty(M_1)$. In particular, if $M_1 = M_2 = M$ and $f =$ id then $\Gamma_{f=id}TM$ is the set of vector fields on M and $\Gamma_{id}TM$ is denoted by ΓTM.

Remark 2.2.1 Let $f : M_1 \to M_2$ be a map. Note that if $X \in \Gamma TM_1$ then the map $f_*X : M_1 \to TM_2$ defined by $(f_*X)(p_1) = f_{*p_1}X(p_1)$ is a vector field along f, where $f_* : TM_1 \to TM_2$ is the tangent map of f. Also, if $Y \in \Gamma TM_2$ then $Y \circ f$ is a vector field along f. Note that the set of all $Y \circ f \in \Gamma_f TM_2$ contains local basis fields for $\Gamma_f TM_2$ over $C^\infty(M_1)$. Also, $X \in \Gamma TM_1$ and $Y \in \Gamma TM_2$ are called *f-related* if $f_*X = Y \circ f$.

Definition 2.2.2 Let $f : M_1 \to M_2$ be a map. A map $\nabla : \Gamma TM_1 \times \Gamma_f TM_2 \to \Gamma_f TM_2$ is called a *connection on M_2 along f* if given X, $Y \in \Gamma TM_1$ and U, $V \in \Gamma_f TM_2$,

(1) $\nabla_{X+Y}U = \nabla_X U + \nabla_Y U$

(2) $\nabla_{hX}U = h\nabla_X U$, $h \in C^\infty(M_1)$

(3) $\nabla_X(U+V) = \nabla_X U + \nabla_X V$

(4) $\nabla_X(hU) = X(h)U + h\nabla_X U$, $h \in C^\infty(M_1)$.

In particular, if $M_1 = M_2 = M$ and $f = \mathrm{id}$ then ∇ is called a *connection on* M.

Remark 2.2.2 Let $f : M_1 \to M_2$ be a map and let ∇ be a connection on M_2 along f. Then since (1) and (2) of Definition 2.2.2 implies the linearity in the first argument of ∇ over $C^\infty(M_1)$, it can be shown as in Proposition 2.1.1 that $(\nabla_X U)(p_1)$ only depends on the value of X at p_1, where $p_1 \in M_1$. Thus we can define $\nabla_x U$ by

$$\nabla_x U = (\nabla_X U)(p_1),$$

where $X \in \Gamma T M_1$ with $X(p_1) = x$.

Definition 2.2.3 Let $f : M_1 \to M_2$ be a map and ∇ be a connection on M_2 along f. Then for each $U \in \Gamma_f T M_2$, the map

$$\nabla U : \Gamma T M_1 \to \Gamma_f T M_2$$

defined by

$$(\nabla U)(X) = \nabla_X U$$

is called the *covariant differential of* U. Also $U \in \Gamma_f T M_2$ is called *parallel* if $\nabla U = 0$.

Theorem 2.2.1 *Let $f : M_1 \to M_2$ be a map and $\overset{2}{\nabla}$ be a connection on M_2. Then there exists a unique connection $\overset{2}{\nabla}{}^f$ on M_2 along f such that, for $Y \in \Gamma T M_2$,*

$$\overset{2}{\nabla}{}^f{}_X\,(Y \circ f) = \overset{2}{\nabla}_{f_* X}\, Y.$$

Proof. Uniqueness: Let $\{Y_1, \ldots, Y_{n_2}\}$ be a local frame for TM_2. Hence, if $Y \in \Gamma_f T M_2$ then locally $Y = \sum_{i=1}^{n_2} h^i (Y_i \circ f)$, where $h^i \in C^\infty(M_1)$, $i = 1, \ldots, n_2$. Thus, if $\overset{2}{\nabla}{}^f$ is such a connection on M_2 along f then, for $X \in \Gamma T M_1$,

$$\begin{aligned} \overset{2}{\nabla}{}^f{}_X\, Y &= \sum_{i=1}^{n_2} X(h^i)(Y_i \circ f) + \sum_{i=1}^{n_2} h^i\, \overset{2}{\nabla}{}^f{}_X\,(Y_i \circ f) \\ &= \sum_{i=1}^{n_2} X(h^i)(Y_i \circ f) + \sum_{i=1}^{n_2} h^i\, \overset{2}{\nabla}_{f_* X}\, Y_i. \end{aligned}$$

Hence $\overset{2}{\nabla}{}^f$ is completely determined by $\overset{2}{\nabla}$, that is, $\overset{2}{\nabla}{}^f$ is unique.

<u>Existence:</u> Define $\overset{2}{\nabla}{}^f$ locally by the above formula. It is then straightforward to check that $\overset{2}{\nabla}{}^f$ is locally a connection on M_2 along f with $\overset{2}{\nabla}{}^f{}_X\,(Y \circ f) = \overset{2}{\nabla}_{f_*X}\,Y$, where $X \in \Gamma TM_1$ and $Y \in \Gamma TM_2$. Hence by uniqueness, these local definitions give rise to a global definition for $\overset{2}{\nabla}{}^f$. □

Definition 2.2.4 Let $f : M_1 \to M_2$ be a map and $\overset{2}{\nabla}$ be a connection on M_2. Then the unique connection $\overset{2}{\nabla}{}^f$ on M_2 along f is called the *pullback of* $\overset{2}{\nabla}$ *along* f. As long as it is clear from the context, we will also denote $\overset{2}{\nabla}{}^f$ by $\overset{2}{\nabla}$.

Let (M, g) be a semi-Riemannian manifold. Next we show that the induced semi-Riemannian structure on M by g defines a canonical connection on M.

Let (M, g) be a semi-Riemannian manifold and let $g^\flat : \Gamma TM \to \Gamma TM^*$ be the musical isomorphism defined by $g^\flat(X) = g(X, \cdot)$. (Note that since g is nondegenerate, $g^\flat$ is injective and hence an isomorphism). Let $g^\sharp : \Gamma TM^* \to \Gamma TM$ be the inverse of $g^\flat$. (See [O, p. 60], [Po, p. 116]).

Theorem 2.2.2 *Let (M, g) be a semi-Riemannian manifold. Then there exists a unique connection ∇ on M with*

(1) $Zg(X,Y) = g(\nabla_Z X, Y) + g(X, \nabla_Z Y)$, X, Y, $Z \in \Gamma TM$,

(2) $\nabla_X Y - \nabla_Y X = [X, Y]$, X, $Y \in \Gamma TM$,

where $[\,,\,]$ *denotes the Lie bracket.*

Proof. <u>Uniqueness:</u> Alternately using the conditions (1) and (2) we see that, for X, Y, $Z \in \Gamma TM$,

$$\begin{aligned}
g(\nabla_X Y, Z) &= Xg(Y,Z) - g(Y, \nabla_X Z) \\
&= Xg(Y,Z) - g(Y, \nabla_Z X + [X,Z]) \\
&= Xg(Y,Z) - Zg(Y,X) + g(\nabla_Z Y, X) + g(Y,[Z,X]) \\
&= Xg(Y,Z) - Zg(X,Y) + g(\nabla_Y Z + [Z,Y], X) \\
&\quad + g(Y,[Z,X]) \\
&= Xg(Y,Z) - Zg(X,Y) + Yg(Z,X) - g(Z, \nabla_Y X) \\
&\quad - g(X,[Y,Z]) + g(Y,[Z,X]) \\
&= Xg(Y,Z) + Yg(Z,X) - Zg(X,Y) + g(Z,[X,Y]) \\
&\quad + g(Y,[Z,X]) - g(X,[Y,Z]) - g(\nabla_X Y, Z).
\end{aligned}$$

Thus

$$\begin{aligned}
g(\nabla_X Y, Z) &= \tfrac{1}{2}[Xg(Y,Z) + Yg(Z,X) - Zg(X,Y) \\
&\quad + g(Z,[X,Y]) + g(Y,[Z,X]) - g(X,[Y,Z])]
\end{aligned}$$

(*Koszul formula*).

Since Koszul formula holds for every connection ∇ on M satisfying the conditions (1) and (2), and since g is nondegenerate, ∇ is unique.

Existence: Fix X, $Y \in \Gamma TM$ and define $\omega_{X,Y} : \Gamma TM \to C^\infty(M)$ by setting $\omega_{X,Y}(Z)$ equal to the right hand side of the Koszul formula for $Z \in \Gamma TM$. It is easy to see that $\omega_{X,Y}$ is linear in its argument. Thus $\omega_{X,Y} \in \Lambda^1(M)$. Now define $\nabla_X Y = g^\sharp(\omega_{X,Y}) \in \Gamma TM$. Then it can easily be checked again that ∇ is a connection on M satisfying the conditions (1) and (2). □

Definition 2.2.5 Let (M,g) be a semi-Riemannian manifold. Then the unique connection ∇ which is given by Theorem 2.2.2 is called the *Levi-Civita connection of* (M,g).

Theorem 2.2.3 *Let M_1 be a manifold and (M_2, g_2) be a semi-Riemannian manifold with Levi-Civita connection $\overset{2}{\nabla}$. If $f : M_1 \to M_2$ is a map then the pullback of $\overset{2}{\nabla}$ along f, which we also denote by $\overset{2}{\nabla}$, satisfies*

(1) $Xg_2(U,W) = g_2(\overset{2}{\nabla}_X U, W) + g_2(U, \overset{2}{\nabla}_X W)$, $X \in \Gamma TM_1$, U, $W \in \Gamma_f TM_2$,

(2) $\overset{2}{\nabla}_X f_*Y - \overset{2}{\nabla}_Y f_*X = f_*[X,Y]$, $X, Y \in \Gamma TM_1$.

Proof. (1) Let $X \in \Gamma TM_1$ be f-related to $X' \in \Gamma TM_2$, that is, $f_*X = X' \circ f$, and let U, $W \in \Gamma TM_2$. Then

$$\begin{aligned} X g_2(U \circ f, W \circ f) &- g_2(\overset{2}{\nabla}_X (U \circ f), W \circ f) - g_2(U \circ f, \overset{2}{\nabla}_X (W \circ f)) \\ &= (f_*X) g_2(U, W) - g_2(\overset{2}{\nabla}_{f_*X} U, W \circ f) - g_2(U \circ f, \overset{2}{\nabla}_{f_*X} W) \\ &= (X' \circ f) g_2(U, W) - g_2(\overset{2}{\nabla}_{X' \circ f} U, W \circ f) - g_2(U \circ f, \overset{2}{\nabla}_{X' \circ f} W) \\ &= (X' g_2(U, W)) \circ f - g_2(\overset{2}{\nabla}_{X'} U, W) \circ f - g_2(U, \overset{2}{\nabla}_{X'} W) \circ f \\ &= (X' g_2(U, W) - g_2(\overset{2}{\nabla}_{X'} U, W) - g_2(U, \overset{2}{\nabla}_{X'} W)) \circ f \\ &= 0. \end{aligned}$$

Now observe that the map

$$T : \Gamma TM_1 \times \Gamma_f TM_2 \times \Gamma_f TM_2 \to C^\infty(M_1)$$

defined by

$$T(X, U, W) = X g_2(U, W) - g_2(\overset{2}{\nabla}_X U, W) - g_2(U, \overset{2}{\nabla}_X W)$$

is linear in its arguments. Hence since the set of all $Y \circ f \in \Gamma_f TM_2$, where $Y \in \Gamma TM_2$, contains local basis fields for $\Gamma_f TM_2$ over $C^\infty(M_1)$, it follows from linearity of T that $T(X, U, W) = 0$ for all $X \in \Gamma TM_1$, U, $W \in \Gamma_f TM_2$.

(2) Let $X, Y \in \Gamma TM_1$ be f-related to X', $Y' \in \Gamma TM_2$, respectively, that is, $f_*X = X' \circ f$ and $f_*Y = Y' \circ f$. Then

$$\overset{2}{\nabla}_X f_*Y = \overset{2}{\nabla}_X (Y' \circ f) = \overset{2}{\nabla}_{f_*X} Y' = \overset{2}{\nabla}_{X' \circ f} Y' = (\overset{2}{\nabla}_{X'} Y') \circ f,$$

and similarly

$$\overset{2}{\nabla}_Y f_*X = (\overset{2}{\nabla}_{Y'} X') \circ f.$$

Also, since X and Y are f-related to X' and Y', respectively,

$$f_*[X, Y] = [X', Y'] \circ f.$$

Thus

$$\overset{2}{\nabla}_X f_*Y - \overset{2}{\nabla}_Y f_*X - f_*[X,Y] = (\overset{2}{\nabla}_{X'} Y' - \overset{2}{\nabla}_{Y'} X' - [X',Y']) \circ f = 0.$$

Now observe that the map

$$T : \Gamma TM_1 \times \Gamma TM_1 \to \Gamma_f TM_2$$

defined by

$$T(X,Y) = \overset{2}{\nabla}_X f_*Y - \overset{2}{\nabla}_Y f_*X - f_*[X,Y]$$

is linear in its arguments. Hence since the set of all $Z \circ f \in \Gamma_f TM_2$, where $Z \in \Gamma TM_2$, contains local basis fields for $\Gamma_f TM_2$ over $C^\infty(M_1)$, it follows from linearity of T that $T(X,Y) = 0$ for all X, $Y \in \Gamma TM_1$. □

Now we study some geometric concepts introduced by the pullback of the Levi-Civita connection along a curve.

Proposition 2.2.1 *Let (M,g) be a semi-Riemannian manifold and let $\gamma : I \to M$ be a curve, where $I \subseteq \mathbf{R}$ is an open interval. Let ∇ be the Levi-Civita connection of (M,g) and let ∇ also denote the pullback of ∇ along γ. If $x \in T_{\gamma(t_0)}M$, where $t_0 \in I$, then there exists a unique $X \in \Gamma_\gamma TM$ with $X(t_0) = x$ and such that $\nabla_{\frac{d}{dt}} X = 0$.*

Proof. Let $\{X_1, \ldots, X_n\}$ be a basis for $\Gamma_\gamma TM$ over $C^\infty(\mathbf{R})$. Also let

$$\nabla_{\frac{d}{dt}} X_i = \sum_{j=1}^n a_i^j X_j,$$

where $a_i^j : I \to \mathbf{R}$, $1 \le i,j \le n$. Now note that, if $X \in \Gamma_\gamma TM$ then $X = \sum\limits_{i=1}^n h^i X_i$, where $h^i : I \to \mathbf{R}$, $1 \le i \le n$. Thus

$$\begin{aligned}
\nabla_{\frac{d}{dt}} X &= \sum_{i=1}^n \frac{dh^i}{dt} X_i + \sum_{i=1}^n h^i \nabla_{\frac{d}{dt}} X_i \\
&= \sum_{i=1}^n \frac{dh^i}{dt} X_i + \sum_{i=1}^n \sum_{j=1}^n h^i a_i^j X_j \\
&= \sum_{j=1}^n \left(\frac{dh^j}{dt} + \sum_{i=1}^n a_i^j h^i \right) X_j.
\end{aligned}$$

Hence, $\nabla_{\frac{d}{dt}} X = 0$ if and only if

$$\frac{dh^j}{dt} + \sum_{i=1}^{n} a_i^j h^i = 0, \quad 1 \leq j \leq n.$$

But this is a first order linear differential system and has a unique solution for a given initial value $(h^1(t_0), \ldots, h^n(t_0))$. Therefore, for a given initial value by $X(t_0) = x = \sum_{i=1}^{n} h^i(t_0) X_i(t_0)$, there is a unique $X \in \Gamma_\gamma TM$ with $X(t_0) = x$ and $\nabla_{\frac{d}{dt}} X = 0$. □

For sake of brevity in terminology, if (M, g) is a semi-Riemannian manifold and $\gamma : I \to M$ is a curve then, $X \in \Gamma_\gamma TM$ is called *parallel along* γ if $\nabla_{\frac{d}{dt}} X = 0$.

Definition 2.2.6 Let (M, g) be a semi-Riemannian manifold and let $\gamma : I \to M$ be a curve with $\gamma(t_1) = p$ and $\gamma(t_2) = q$, where $I \subseteq \mathbf{R}$ is an open interval and $t_1, t_2 \in I$. Then the map

$$P_{t_1}^{t_2}(\gamma) : T_pM \to T_qM$$

defined by $P_{t_1}^{t_2}(\gamma)(x) = z$, where $z = X(t_2) \in T_qM$ and X is the parallel vector field along γ with $X(t_1) = x$, is called *parallel translation along* γ *from* p *to* q.

Proposition 2.2.2 *Let* (M, g) *be a semi-Riemannian manifold and let* $\gamma : I \to M$ *be a curve. Then the parallel translation* $P_{t_1}^{t_2}(\gamma) : (T_pM, g_p) \to (T_qM, g_q)$ *is a linear isometry.*

Proof. With notation as in Definition 2.2.6, let $x, y \in T_pM$ and $X, Y \in \Gamma_\gamma TM$ be parallel vector fields along γ with $X(t_1) = x$, $Y(t_1) = y$. Since $X + Y$ is also parallel with $(X + Y)(t_0) = X(t_0) + Y(t_0) = x + y$,

$$P_{t_1}^{t_2}(\gamma)(x + y) = (X + Y)(t_2) = X(t_2) + Y(t_2) = P_{t_1}^{t_2}(\gamma)(x) + P_{t_1}^{t_2}(\gamma)(y).$$

Similarly

$$P_{t_1}^{t_2}(\gamma)(cx) = cP_{t_1}^{t_2}(\gamma)(x),$$

where $c \in \mathbf{R}$. Thus $P_{t_1}^{t_2}(\gamma)$ is linear. If $P_{t_1}^{t_2}(\gamma)(x) = 0$, then by the uniqueness in Proposition 2.2.1, X can only be the zero vector field along γ. Hence $x = X(t_0) = 0$. Thus $P_{t_1}^{t_2}(\gamma)$ is one-to-one, and since dim T_pM = dim T_qM, $P_{t_1}^{t_2}(\gamma)$ is a linear isomorphism. Finally, for X, Y as in the above, it follows from Theorem 2.2.3 that

$$\frac{d}{dt}g(X,Y) = g(\nabla_{\frac{d}{dt}}X,Y) + g(X,\nabla_{\frac{d}{dt}}Y) = 0.$$

Thus $g(X,Y)$ is constant along γ and hence

$$g(P_{t_1}^{t_2}(\gamma)(x), P_{t_1}^{t_2}(\gamma)(y)) = g(X(t_2),Y(t_2)) = g(X(t_1),Y(t_1)) = g(x,y).$$

That is, $P_{t_1}^{t_2}(\gamma)$ is an isometry. □

Let M be a manifold and $\gamma : I \to M$ be a curve, where $I \subseteq \mathbf{R}$ is an open interval. Then the *velocity vector field* of γ is defined by $\dot{\gamma} = \gamma_* \frac{d}{dt}$. Note that $\dot{\gamma}$ is a vector field along γ.

Definition 2.2.7 Let (M,g) be a semi-Riemannian manifold and $\gamma : I \to M$ be a curve, where $I \subseteq \mathbf{R}$ is an open interval. Then γ is called a *geodesic of* (M,g) if $\dot{\gamma} \in \Gamma_\gamma TM$ is parallel along γ, that is, $\nabla_{\frac{d}{dt}} \dot{\gamma} = 0$.

Theorem 2.2.4 *Let (M,g) be a semi-Riemannian manifold. Then for each $v \in T_pM$, there exists a geodesic $\gamma : I \to M$ of (M,g) with $\dot{\gamma}(t_0) = v$, where $t_0 \in I$ and $I \subseteq \mathbf{R}$ is an open interval. Furthermore, γ is unique up to domain of definition.*

Proof. Let $\varphi = (x^1,\ldots,x^n)$ be a chart whose domain contains $p \in M$. Hence, if $\gamma : I \to M$ is a curve with $\gamma(t_0) = p$ then $\dot{\gamma} = \sum_{i=1}^n \frac{d\gamma^i}{dt}\frac{\partial}{\partial x^i} \circ \gamma$, where $\gamma^i = x^i \circ \gamma$. Thus

$$\begin{aligned}
\nabla_{\frac{d}{dt}} \dot{\gamma} &= \sum_{i=1}^n \frac{d^2\gamma^i}{dt^2}(\frac{\partial}{\partial x^i} \circ \gamma) + \sum_{i=1}^n \frac{d\gamma^i}{dt}\nabla_{\frac{d}{dt}}(\frac{\partial}{\partial x^i} \circ \gamma) \\
&= \sum_{i=1}^n \frac{d^2\gamma^i}{dt^2}(\frac{\partial}{\partial x^i} \circ \gamma) + \sum_{i,j=1}^n \frac{d\gamma^i}{dt}\frac{d\gamma^j}{dt}\nabla_{\frac{\partial}{\partial x^j}\circ\gamma}\frac{\partial}{\partial x^i} \\
&= \sum_{i=1}^n \frac{d^2\gamma^i}{dt^2}(\frac{\partial}{\partial x^i} \circ \gamma) \\
&\quad + \sum_{i,j,k=1}^n \frac{d\gamma^i}{dt}\frac{d\gamma^j}{dt}\Gamma_{ji}^k \circ \varphi^{-1}(\gamma^1,\ldots,\gamma^n)(\frac{\partial}{\partial x^k} \circ \gamma) \\
&= \sum_{k=1}^n \left[\frac{d^2\gamma^k}{dt^2} + \sum_{i,j=1}^n \frac{d\gamma^i}{dt}\frac{d\gamma^j}{dt}\Gamma_{ji}^k \circ \varphi^{-1}(\gamma^1,\ldots,\gamma^n)\right](\frac{\partial}{\partial x^k} \circ \gamma),
\end{aligned}$$

where $\nabla_{\frac{\partial}{\partial x^j}} \frac{\partial}{\partial x^i} = \sum_{k=1}^{n} \Gamma_{ji}^k \frac{\partial}{\partial x^k}$.

Hence, the equation for a geodesic is the second order differential system given by

$$\frac{d^2\gamma^k}{dt^2} + \sum_{i,j=1}^{n} \frac{d\gamma^i}{dt}\frac{d\gamma^j}{dt}\Gamma_{ji}^k \circ \varphi^{-1}(\gamma^1, \ldots, \gamma^n) = 0, \qquad k = 1, \ldots, n,$$

with initial conditions

$$(\gamma^1(t_0), \ldots, \gamma^n(t_0)) = \varphi(p)$$

and

$$\left(\frac{d\gamma^1}{dt}(t_0), \ldots, \frac{d\gamma^n}{dt}(t_0)\right) = (dx^1(v), \ldots, dx^n(v)).$$

For such a system, existence, uniqueness and C^∞ dependence on initial conditions are all guaranteed by standard theorems on differential equations; existence is guaranteed only locally. □

Proposition 2.2.3 *Let (M, g) be a semi-Riemannian manifold and $\gamma : (a, b) \to M$ be a nonconstant geodesic of (M, g). If $h : (c, d) \to (a, b)$ is a function then, the curve $\alpha = \gamma \circ h$ is a geodesic of (M, g) if and only if $t = h(s) = As + B$, where A, $B \in \mathbf{R}$.*

Proof. Let ∇ denote both pullbacks of the Levi-Civita connection of (M, g) along the curves α and γ. Then since $\dot{\alpha} = \frac{dh}{ds}(\dot{\gamma} \circ h)$,

$$\begin{aligned}
\nabla_{\frac{d}{ds}} \dot{\alpha} &= \frac{d^2h}{ds^2}(\dot{\gamma} \circ h) + \frac{dh}{ds}\nabla_{\frac{d}{ds}}(\dot{\gamma} \circ h) \\
&= \frac{d^2h}{ds^2}(\dot{\gamma} \circ h) + \frac{dh}{ds}\nabla_{h_* \frac{d}{ds}} \dot{\gamma} \\
&= \frac{d^2h}{ds^2}(\dot{\gamma} \circ h) + \frac{dh}{ds}\nabla_{\frac{dh}{ds}\frac{d}{dt} \circ h} \dot{\gamma} \\
&= \frac{d^2h}{ds^2}(\dot{\gamma} \circ h) + \left(\frac{dh}{ds}\right)^2 \nabla_{\frac{d}{dt} \circ h} \dot{\gamma} \\
&= \frac{d^2h}{ds^2}(\dot{\gamma} \circ h) + \left(\frac{dh}{ds}\right)^2 (\nabla_{\frac{d}{dt}} \dot{\gamma}) \circ h \\
&= \frac{d^2h}{ds^2}(\dot{\gamma} \circ h)
\end{aligned}$$

since γ is a geodesic of (M, g).

Hence, since $\dot{\gamma} \circ h \neq 0$, it follows that, $\frac{d^2h}{ds^2} = 0$ if and only if α is a geodesic of (M, g). □

It may be impossible to extend a geodesic $\gamma : I \to M$ of a semi-Riemannian manifold (M, g) to a geodesic defined on all of $\mathbf{R}$, where $I \subseteq \mathbf{R}$ is an open interval. We call a geodesic $\gamma : I \to M$ of (M, g) *inextendable*, where $I \subseteq \mathbf{R}$ is an open interval, if γ cannot be extended to a geodesic of (M, g) with larger interval of definition.

Definition 2.2.8 *A semi-Riemannian manifold (M, g) is called geodesically complete if every inextendable geodesic of (M, g) has domain $\mathbf{R}$.*

At each point p of a semi-Riemannian manifold (M, g), we collect the geodesics starting at p into a single mapping.

Let (M, g) be a semi-Riemannian manifold. We denote the geodesic $\gamma : I \to M$ of (M, g) with $\dot{\gamma}(0) = v \in T_pM$ by γ_v, where $I \subseteq \mathbf{R}$ is an open interval with $0 \in I$.

Definition 2.2.9 Let (M, g) be a semi-Riemannian manifold. Let $p \in M$ and $\mathcal{D}_p$ be the set of vectors $v \in T_pM$ such that the inextendable geodesic γ_v is defined at least on $[0, 1]$. Then the *exponential map of (M, g) at $p \in M$* is the map

$$\exp_p : \mathcal{D}_p \to M$$

such that

$$\exp_p(v) = \gamma_v(1)$$

for all $v \in \mathcal{D}_p$.

Note that if $v \in T_pM$ and $t \in \mathbf{R}$ then the geodesic $\beta(s) = \gamma_v(ts)$ (cf. Proposition 2.2.3) has initial velocity $\dot{\beta}(0) = tv$. Hence $\gamma_{tv}(s) = \gamma_v(ts)$ for all s and t whenever both sides are defined. Thus, if $v \in \mathcal{D}_p$ then

$$\exp_p(tv) = \gamma_{tv}(1) = \gamma_v(t).$$

Theorem 2.2.5 *Let (M, g) be a semi-Riemannian manifold and let $\exp_p : \mathcal{D}_p \to M$ be the exponential map of (M, g) at $p \in M$. Then there exists a neighborhood $\tilde{\mathcal{U}}$ of $0 \in T_pM$ on which $\exp_p$ is a diffeomorphism onto a neighborhood $\mathcal{U}$ of p in M.*

Proof. It is well-known from the theory of differential equations that, if a system of second order ordinary differential equations is given by smooth functions, then its solutions are smooth in the parameter, initial values and initial first derivatives. Thus, applying this fact to the definition $\exp_p$, it yields smoothness of $\exp_p$ on some neighborhood of $0 \in T_pM$. To show that $\exp_p$ is a diffeomorphism from a neighborhood $\tilde{\mathcal{U}}$ of $0 \in T_pM$ onto a neighborhood $\mathcal{U}$ of $p \in M$, it suffices to show that $\exp_{p_{*0}} : T_0T_pM \to T_pM$ is nonsingular. For this, let $c : I \to T_pM$ be a curve defined by $c(t) = tv$ on some sufficiently small interval I, where $v \in T_pM$. Then note that $\dot{c}\,(0) = v_0 = v$ under the canonical identification of T_0T_pM and T_pM. Hence, by $\exp_p(tv) = \gamma_v(t)$ noted above,

$$\begin{aligned}
\exp_{p_{*0}}(v_0) &= \exp_{p_{*0}}(\dot{c}\,(0)) \\
&= (\exp_p \circ c)_{*0}\tfrac{d}{dt}\,|_{t=0} \\
&= \gamma_{v*0}\tfrac{d}{dt}\,|_{t=0} \\
&= \dot{\gamma_v}\,(0) = v.
\end{aligned}$$

Hence $\exp_p$ is nonsingular at $0 \in T_pM$ and the result then follows by the inverse function theorem. □

Let (M, g) be a semi-Riemannian manifold. Then the exponential maps $\exp_p : \mathcal{D}_p \to M$ of (M, g) for all $p \in M$ also constitute a single mapping exp: $\mathcal{D} \to M$, where $\mathcal{D} = \bigcup_{p\in M} \mathcal{D}_p$, by defining

$$\exp(v) = \exp_p(v),$$

where $v \in \mathcal{D}_p \subseteq T_pM$. We call exp: $\mathcal{D} \to M$ the *exponential map of* (M, g). It is clear that the exponential map of (M, g) is a smooth function.

Finally we construct a local basis frame on a neighborhood of a point of a semi-Riemannian manifold which will be useful in computations throughout this book. Let (M, g) be a semi-Riemannian manifold and $p \in M$. Note that, by Theorem 2.2.5, there is a neighborhood $\mathcal{U}$ of p such that the geodesics emanating from p do not intersect each other on $\mathcal{U}$. Let $\{x_1, \ldots, x_n\}$ be an orthonormal basis for (T_pM, g_p) and let $X_1, \ldots, X_n$ be vector fields on $\mathcal{U}$ obtained by parallel translation of $x_1, \ldots, x_n$ along the geodesics emanating from p, respectively. By Proposition 2.2.2, $X_1, \ldots, X_n$ are orthonormal vector fields on $\mathcal{U}$ which form a basis for (T_qM, g_q) at each $q \in \mathcal{U}$. We call such an $\{X_1, \ldots, X_n\}$ an *adapted moving frame near p*. Note that the most important property of an adapted moving frame $\{X_1, \ldots, X_n\}$ near $p \in M$ is $(\nabla X_i)_p = 0$ for $i = 1, \ldots, n$.

2.3 Curvatures of Semi-Riemannian Manifolds

In this section, we define some geometric objects which are used in the description of the geometry of semi-Riemannian manifolds. Again let M, M_1 and M_2 denote manifolds of dimensions n, n_1 and n_2, respectively. As usual, also let (M, g) denote a semi-Riemannian manifold (of index ν).

Definition 2.3.1 Let $f : M_1 \to M_2$ be a map and ∇ be a connection on M_2 along f. Then the *curvature tensor* R *of* ∇ is defined by

$$R(X,Y)U = \nabla_X\nabla_Y U - \nabla_Y\nabla_X U - \nabla_{[X,Y]}U,$$

where X, $Y \in \Gamma TM_1$, $U \in \Gamma_f TM_2$ and $[\,,\,]$ is the Lie bracket.

It is easy to check that the map

$$R : \Gamma TM_1 \times \Gamma TM_1 \times \Gamma_f TM_2 \to \Gamma_f TM_2$$

defined by

$$R(X,Y,U) = R(X,Y)U$$

is linear in its arguments. For example, if $h \in C^\infty(M_1)$,

$$\begin{aligned}
R(X,Y)(hU) &= \nabla_X\nabla_Y(hU) - \nabla_Y\nabla_X(hU) - \nabla_{[X,Y]}(hU) \\
&= \nabla_X(Y(h)U + h\nabla_Y U) - \nabla_Y(X(h)U + h\nabla_X U) \\
&\quad -[X,Y](h)U - h\nabla_{[X,Y]}U \\
&= X(Y(h))U + Y(h)\nabla_X U + X(h)\nabla_Y U \\
&\quad +h\nabla_X\nabla_Y U - Y(X(h))U - X(h)\nabla_Y U - Y(h)\nabla_X U \\
&\quad -h\nabla_Y\nabla_X U - X(Y(h))U + Y(X(h))U - h\nabla_{[X,Y]}U \\
&= hR(X,Y)U
\end{aligned}$$

Note that, because of the linearity of R in its arguments, it can be shown as in Proposition 2.1.1 that the value of $R(X,Y)U$ at $p_1 \in M_1$ only depends on the values of X, Y and U at p_1. Hence one can define $R(x,y)u$ by

$$R(x,y)u = (R(X,Y)U)(p_1),$$

where X, $Y \in \Gamma TM_1$, $U \in \Gamma_f TM_2$ with $X(p_1) = x$, $Y(p_1) = y$, $U(p_1) = u$.

Proposition 2.3.1 *Let $f : M_1 \to M_2$ be a map and $\overset{2}{\nabla}$ be a connection on M_2 with curvature tensor $\overset{2}{R}$. Also let $\overset{2}{\nabla}{}^f$ be the pullback of $\overset{2}{\nabla}$ along f with curvature tensor $\overset{2}{R}{}^f$. Then for X, $Y \in \Gamma TM_1$, $U \in \Gamma_f TM_2$,*

$$\overset{2}{R}(f_*X, f_*Y)U = \overset{2}{R}{}^f(X,Y)U.$$

Proof. Let X, $Y \in \Gamma TM_1$ be f-related to X', $Y' \in \Gamma TM_2$, respectively, that is, $f_*X = X' \circ f$ and $f_*Y = Y' \circ f$. Also let $U = U' \circ f$, where $U' \in \Gamma TM_2$. Then

$$\begin{aligned}
\overset{2}{\nabla}{}^f{}_X \overset{2}{\nabla}{}^f{}_Y U &= \overset{2}{\nabla}{}^f{}_X \overset{2}{\nabla}{}^f{}_Y (U' \circ f) \\
&= \overset{2}{\nabla}{}^f{}_X \overset{2}{\nabla}_{f_*Y} U' = \overset{2}{\nabla}{}^f{}_X \overset{2}{\nabla}_{Y'\circ f} U' \\
&= \overset{2}{\nabla}{}^f{}_X ((\overset{2}{\nabla}_{Y'} U') \circ f) = \overset{2}{\nabla}_{f_*X} \overset{2}{\nabla}_{Y'} U' \\
&= \overset{2}{\nabla}_{X'\circ f} \overset{2}{\nabla}_{Y'} U' = (\overset{2}{\nabla}_{X'} \overset{2}{\nabla}_{Y'} U') \circ f.
\end{aligned}$$

Similarly

$$\overset{2}{\nabla}{}^f{}_Y \overset{2}{\nabla}{}^f{}_X U = (\overset{2}{\nabla}_{Y'} \overset{2}{\nabla}_{X'} U') \circ f \quad \text{and} \quad \overset{2}{\nabla}{}^f{}_{[X,Y]} U = (\overset{2}{\nabla}_{[X',Y']} U') \circ f.$$

Thus

$$\begin{aligned}
\overset{2}{R}{}^f(X,Y)U &= \overset{2}{\nabla}{}^f{}_X \overset{2}{\nabla}{}^f{}_Y U - \overset{2}{\nabla}{}^f{}_Y \overset{2}{\nabla}{}^f{}_X U - \overset{2}{\nabla}{}^f{}_{[X,Y]} U \\
&= \left(\overset{2}{\nabla}_{X'} \overset{2}{\nabla}_{Y'} U' - \overset{2}{\nabla}_{Y'} \overset{2}{\nabla}_{X'} U' - \overset{2}{\nabla}_{[X',Y']} U'\right) \circ f \\
&= \left(\overset{2}{R}(X', Y')U'\right) \circ f \\
&= \overset{2}{R}(X' \circ f, Y' \circ f)U' \circ f \\
&= \overset{2}{R}(f_*X, f_*Y)U.
\end{aligned}$$

Note that the set of all $Z \circ f$, where $Z \in \Gamma TM_2$, contains local basis fields for $\Gamma_f TM_2$ over $C^\infty(M_1)$. Hence it follows from linearity of $\overset{2}{R}{}^f$ and $\overset{2}{R}$ that $\overset{2}{R}{}^f(X,Y)U = \overset{2}{R}(f_*X, f_*Y)U$ for all X, $Y \in \Gamma TM_1$ and $U \in \Gamma_f TM_2$. □

As long as it is clear from the context, we will also denote $\overset{2}{R^f}$ by $\overset{2}{R}$.

Now we state some symmetry properties of the curvature tensor of the Levi-Civita connection of a semi-Riemannian manifold.

Proposition 2.3.2 *Let ∇ be the Levi-Civita connection of a semi-Riemannian manifold (M,g) and let R be the curvature tensor of ∇. Then for all X, Y, Z, $V \in \Gamma TM$,*

(1) $R(X,Y)Z = -R(Y,X)Z$

(2) $g(R(X,Y)Z,V) = -g(R(X,Y)V,Z)$

(3) $R(X,Y)Z + R(Y,Z)X + R(Z,X)Y = 0$

(4) $g(R(X,Y)Z,V) = g(R(Z,V)X,Y)$

Proof. (1) Obvious from Definition 2.3.1.

(2) It suffices to show that $g(R(X,Y)Z,Z) = 0$ for all $Z \in \Gamma TM$. Indeed

$$\begin{aligned}
g(R(X,Y)Z,Z) &= g(\nabla_X\nabla_Y Z,Z) - g(\nabla_Y\nabla_X Z,Z) - g(\nabla_{[X,Y]}Z,Z) \\
&= Xg(\nabla_Y Z,Z) - g(\nabla_Y Z,\nabla_X Z) \\
&\quad -Yg(\nabla_X Z,Z) + g(\nabla_X Z,\nabla_Y Z) \\
&\quad -\tfrac{1}{2}[X,Y]g(Z,Z) \\
&= \tfrac{1}{2}XYg(Z,Z) - \tfrac{1}{2}YXg(Z,Z) - \tfrac{1}{2}[X,Y]g(Z,Z) \\
&= 0
\end{aligned}$$

(3)

$$
\begin{aligned}
&R(X,Y)Z + R(Y,Z)X + R(Z,X)Y \\
&\quad = \nabla_X\nabla_Y Z - \nabla_Y\nabla_X Z - \nabla_{[X,Y]}Z \\
&\qquad +\nabla_Y\nabla_Z X - \nabla_Z\nabla_Y X - \nabla_{[Y,Z]}X \\
&\qquad +\nabla_Z\nabla_X Y - \nabla_X\nabla_Z Y - \nabla_{[Z,X]}Y \\
&\quad = \nabla_X(\nabla_Y Z - \nabla_Z Y) + \nabla_Y(\nabla_Z X - \nabla_X Z) \\
&\qquad +\nabla_Z(\nabla_X Y - \nabla_Y X) - \nabla_{[X,Y]}Z - \nabla_{[Y,Z]}X - \nabla_{[Z,X]}Y \\
&\quad = \nabla_X[Y,Z] + \nabla_Y[Z,X] + \nabla_Z[X,Y] \\
&\qquad -\nabla_{[X,Y]}Z - \nabla_{[Y,Z]}X - \nabla_{[Z,X]}Y \\
&\quad = [X,[Y,Z]] + [Y,[Z,X]] + [Z,[X,Y]] \\
&\quad = 0,
\end{aligned}
$$

where the last equality follows from the Jacobi identity for vector fields.

(4) By using (1), (2) and (3),

$$
\begin{aligned}
g(R(X,Y)Z,V) &= -g(R(X,Y)V,Z) \\
&= g(R(Y,V)X,Z) + g(R(V,X)Y,Z) \\
&= -g(R(Y,V)Z,X) - g(R(V,X)Z,Y) \\
&= g(R(V,Z)Y,X) + g(R(Z,Y)V,X) \\
&\quad +g(R(X,Z)V,Y) + g(R(Z,V)X,Y) \\
&= 2g(R(Z,V)X,Y) + g(R(Y,Z)X,V) + g(R(Z,X)Y,V) \\
&= 2g(R(Z,V)X,Y) - g(R(X,Y)Z,V).
\end{aligned}
$$

Hence $g(R(X,Y)Z,V) = g(R(Z,V)X,Y)$. □

Definition 2.3.2 Let (M,g) be a semi-Riemannian manifold. Then the curvature tensor of the Levi-Civita connection of (M,g) is called the *curvature tensor of* (M,g).

Remark 2.3.1 Note here that if (M,g) is a 1-dimensional semi-Riemannian manifold then the curvature tensor R of (M,g) vanishes.

Curvature tensor of a semi-Riemannian manifold (M, g) also determines other geometric objects on (M, g) which are more meaningful from the geometric viewpoint.

Let (M, g) be a semi-Riemannian manifold. A 2-dimensional subspace P of T_pM is called a *plane in* T_pM. In general, the restriction of g_p to a plane P may be either of type $(0,2,0)$, $(0,0,2)$, $(0,1,1)$, $(2,0,0)$, $(1,1,0)$ or $(1,0,1)$. Hence we call a plane P in T_pM of *signature* $(-,-)$, $(+,+)$, $(-,+)$, $(0,0)$, $(0,-)$ or $(0,+)$ if the restriction of g_p to P is of type $(0,2,0)$, $(0,0,2)$, $(0,1,1)$, $(2,0,0)$, $(1,1,0)$ or $(1,0,1)$, respectively. We call the planes in T_pM of signatures $(-,-)$, $(+,+)$ or $(-,+)$ *nondegenerate* since the restrictions of g_p to such planes are nondegenerate. Also we call the planes in T_pM of signatures $(0,0)$, $(0,-)$ or $(0,+)$ *degenerate* since the restrictions of g_p to such planes are degenerate.

Now let (M, g) be a semi-Riemannian manifold. For any pair of vectors x, $y \in T_pM$, define

$$Q(x, y) = g(x, x)g(y, y) - g(x, y)^2.$$

Lemma 2.3.1 *Let (M, g) be a semi-Riemannian manifold and let P be a plane in T_pM. Then P is nondegenerate if and only if $Q(x, y) \neq 0$, where $P = span\{x, y\}$. Furthermore, P is of signature $(-,-)$ or $(+,+)$ if and only if $Q(x, y) > 0$, and P is of signature $(-,+)$ if and only if $Q(x, y) < 0$.*

Proof. First we show that the sign of Q is independent of the choice of basis for P. Indeed, let $\{x, y\}$ and $\{x', y'\}$ be two bases for P. Then $x' = ax + by$ and $y' = cx + dy$ with $ad - bc \neq 0$. Hence by a direct computation, it can be shown that $Q(x', y') = (ad - bc)^2 Q(x, y)$. Thus the sign of Q is independent of the choice of basis for P. Now suppose P is nondegenerate and let $\{x, y\}$ be an orthonormal basis for (P, g_{pP}). Then it follows that $Q(x, y) = g(x, x)g(y, y) \neq 0$. In particular, P is either of signatures $(-,-)$ or $(+,+)$ if and only if $Q(x, y) > 0$, and P is of signature $(-,+)$ if and only if $Q(x, y) < 0$. Conversely, suppose P is degenerate and let $\{x, y\}$ be an orthonormal basis for P with respect to g_{pP}. (Hence, either x or y is a null vector). Then it follows that $Q(x, y) = 0$. □

Definition 2.3.3 Let (M, g) be a semi-Riemannian manifold and let R be the curvature tensor of (M, g). Then the *sectional curvature* of a nondegenerate plane P in T_pM is defined by

$$K(P) = \frac{g(R(x, y)y, x)}{Q(x, y)},$$

where $P = \text{span}\{x, y\}$.

The sectional curvature of a nondegenerate plane is well-defined, that is, independent of the choice of the basis for P. To see this, let $\{x, y\}$ and $\{x', y'\}$ be two bases for P. Then $x' = ax + by$ and $y' = cx + dy$, where $ad - bc \neq 0$. Hence by a direct computation, it can be shown that

$$g(R(x', y')y', x') = (ad - bc)^2 g(R(x, y)y, x)$$

and

$$Q(x', y') = (ad - bc)^2 Q(x, y).$$

Thus,

$$K(P) = \frac{g(R(x', y')y', x')}{Q(x', y')} = \frac{g(R(x, y)y, x)}{Q(x, y)},$$

that is, the sectional curvature of a nondegenerate plane is well-defined.

Let (M, g) be a semi-Riemannian manifold and let R be the curvature tensor of (M, g). Then it follows from the linearity of R in its arguments that, for fixed Y, $Z \in \Gamma TM$, the map

$$R(\cdot, Y)Z = (X \mapsto R(X, Y)Z)$$

is a linear map from ΓTM to ΓTM, for which the trace is well-defined.

Definition 2.3.4 Let (M, g) be a semi-Riemannian manifold and let R be the curvature tensor of (M, g). Then the map

$$Ric : \Gamma TM \times \Gamma TM \to C^\infty(M)$$

defined by

$$Ric(Y, Z) = \text{trace } R(\cdot, Y)Z$$

is called the *Ricci tensor of* (M, g).

Note that, clearly, the Ricci tensor of (M, g) is linear in its arguments and hence Ric is a $(0, 2)$-tensor field on (M, g).

Proposition 2.3.3 *Let (M, g) be a semi-Riemannian manifold and let Ric be the Ricci tensor of (M, g). Then $Ric(Y, Z) = Ric(Z, Y)$ for every Y, $Z \in \Gamma TM$, that is, the Ricci tensor of (M, g) is symmetric.*

Proof. Let $\{X_1, \ldots, X_n\}$ be an orthonormal local frame. Then by Proposition 2.3.2,

$$\begin{aligned}
Ric(Y, Z) &= \operatorname{trace} R(\cdot\,, Y)Z \\
&= \sum_{i=1}^{n} g(X_i, X_i) g(R(X_i, Y)Z, X_i) \\
&= \sum_{i=1}^{n} g(X_i, X_i) g(R(Y, X_i)X_i, Z) \\
&= \sum_{i=1}^{n} g(X_i, X_i) g(R(X_i, Z)Y, X_i) \\
&= \operatorname{trace} R(\cdot\,, Z)Y \\
&= Ric(Z, Y).
\end{aligned}$$

□

Definition 2.3.5 Let (M, g) be a semi-Riemannian manifold and let Ric be the Ricci tensor of (M, g). The *Ricci curvature* of (M, g) with respect to a nonzero, nonnull vector $v \in TM$ is defined by

$$r(v) = \frac{Ric(v, v)}{|\, g(v, v)\,|}.$$

Also the *scalar curvature* of (M, g) is defined to be the trace of Ric with respect to g, that is,

$$S = \sum_{i=1}^{n} g(X_i, X_i) Ric(X_i, X_i)$$

where $\{X_1, \ldots, X_n\}$ is an orthonormal local frame.

Remark 2.3.2 Note that, if $\{X_1, \ldots, X_n\}$ is an orthonormal local frame,

$$\begin{aligned} S &= \sum_{i=1}^{n} g(X_i, X_i) Ric(X_i, X_i) \\ &= \sum_{i,j=1}^{n} g(X_i, X_i) g(X_j, X_j) g(R(X_j, X_i)X_i, X_j) \\ &= \sum_{i \neq j}^{n} K(P_{ij}) \\ &= 2\sum_{i<j}^{n} K(P_{ij}), \end{aligned}$$

where $P_{ij} = \text{span}\{X_i, X_j\}$, $i \neq j$, are (nondegenerate) planes spanned by X_i and X_j at each point.

2.4 Some Differential Operators

In this section, we define some differential operators on a semi-Riemannian manifold which will be encountered in the study of semi-Riemannian maps.

Throughout this section let (M, g) be an n-dimensional semi-Riemannian manifold and ∇ be the Levi-Civita connection of (M, g).

Recall that, if $Z \in \Gamma TM$, then $\nabla Z : \Gamma TM \to \Gamma TM$ is a linear map, for which the trace is well-defined. (See Definition 2.2.3).

Definition 2.4.1 Let $Z \in \Gamma TM$ be a vector field on a semi-Riemannian manifold (M, g). Then the *divergence* of Z on (M, g) is defined by

$$\text{div } Z = \text{trace } \nabla Z.$$

Note that, if $\{X_1, \ldots, X_n\}$ is an orthonormal local frame then

$$\text{div } Z = \sum_{i=1}^{n} g(X_i, X_i) g(\nabla_{X_i} Z, X_i).$$

Definition 2.4.2 Let $f : (M, g) \to \mathbf{R}$ be a map. Then the *gradient* of f on (M, g) is defined by

$$\nabla f = g^{\sharp}(df),$$

where $g^{\sharp} : \Gamma TM^* \to \Gamma TM$ is the inverse of the musical isomorphism $g^{\flat} : \Gamma TM \to \Gamma TM^*$ defined by $g^{\flat}(X) = g(X, \cdot)$.

Note that, by definition, ∇f is the unique vector field on M determined by

$$g(\nabla f, X) = df(X) = X(f),$$

where $X \in \Gamma TM$.

Definition 2.4.3 Let $f : (M, g) \to \mathbf{R}$ be a map. Then the linear map

$$h_f : \Gamma TM \to \Gamma TM$$

defined by

$$h_f(X) = \nabla_X \nabla f$$

is called the *Hessian tensor of f on (M, g)*.

Proposition 2.4.1 *Let h_f be the Hessian tensor of a map $f : (M, g) \to \mathbf{R}$ on (M, g). Then $g(h_f(X), Y) = g(X, h_f(Y))$ for every X, $Y \in \Gamma TM$, that is, h_f is self-adjoint.*

Proof. Let X, $Y \in \Gamma TM$. Then

$$\begin{aligned}
g(h_f(X), Y) &= g(\nabla_X \nabla f, Y) \\
&= Xg(\nabla f, Y) - g(\nabla f, \nabla_X Y) \\
&= Xg(\nabla f, Y) - g(\nabla f, \nabla_Y X) - g(\nabla f, [X, Y]) \\
&= Xg(\nabla f, Y) - Yg(\nabla f, X) \\
&\quad + g(\nabla_Y \nabla f, X) - g(\nabla f, [X, Y]) \\
&= XY(f) - YX(f) + g(\nabla_Y \nabla f, X) - [X, Y](f) \\
&= [X, Y](f) + g(\nabla_Y \nabla f, X) - [X, Y](f) \\
&= g(\nabla_Y \nabla f, X) \\
&= g(h_f(Y), X).
\end{aligned}$$

□

Definition 2.4.4 Let $f : (M, g) \to \mathbf{R}$ be a map. Then the linear map

$$H_f : \Gamma TM \times \Gamma TM \to C^\infty(M)$$

defined by

$$H_f(X,Y) = g(h_f(X),Y)$$

is called the *Hessian form of f on (M,g)*.

Note that, since the Hessian tensor of f is self-adjoint, the Hessian form of f is a symmetric $(0,2)$-tensor field on (M,g).

Definition 2.4.5 Let $f : (M,g) \to \mathbf{R}$ be a map. The *Laplacian of f on (M,g)* is defined by

$$\Delta f = -\operatorname{div} \nabla f.$$

Note that, Δf is the negative of the trace of h_f. Indeed, if $\{X_1,\ldots,X_n\}$ is an orthonormal local frame,

$$\begin{aligned} \Delta f &= -\operatorname{div} \nabla f \\ &= -\sum_{i=1}^{n} g(X_i,X_i) g(\nabla_{X_i}\nabla f, X_i) \\ &= -\sum_{i=1}^{n} g(X_i,X_i) g(h_f(X_i),X_i) \\ &= -\operatorname{trace} h_f. \end{aligned}$$

Definition 2.4.6 Let $f : (M,g) \to \mathbf{R}$ be a map. Then f is called *harmonic* on (M,g) if $\Delta f = 0$, and f is called *subharmonic* on (M,g) if $-\Delta f \geq 0$.

Proposition 2.4.2 *Let $f,k : (M,g) \to \mathbf{R}$ be maps and X, $Y \in \Gamma TM$. Then*

(1) *$div\,(fX) = g(\nabla f, X) + f div\, X$.*

(2) *$\nabla(fk) = f\nabla k + k\nabla f$.*

(3) *$h_{fk}(X) = fh_k(X) + kh_f(X) + g(\nabla f, X)\nabla k + g(\nabla k, X)\nabla f$.*

(4) *$H_{fk}(X,Y) = fH_k(X,Y) + kH_f(X,Y)$*
$\qquad + g(\nabla f, X)g(\nabla k, Y) + g(\nabla k, X)g(\nabla f, Y)$

(5) *$\Delta(fk) = f\Delta k + k\Delta f - 2g(\nabla f, \nabla k)$.*

Proof. (1) Let $\{X_1, \ldots, X_n\}$ be an orthonormal local frame. Then

$$\begin{aligned}
\text{div } (fX) &= \sum_{i=1}^{n} g(X_i, X_i) g(\nabla_{X_i}(fX), X_i) \\
&= \sum_{i=1}^{n} g(X_i, X_i) g(X_i(f)X, X_i) + \sum_{i=1}^{n} g(X_i, X_i) g(f\nabla_{X_i}X, X_i) \\
&= \sum_{i=1}^{n} g(X_i, X_i) g(\nabla f, X_i) g(X, X_i) + f \sum_{i=1}^{n} g(X_i, X_i) g(\nabla_{X_i}X, X_i) \\
&= g(\nabla f, X) + f \text{ div } X.
\end{aligned}$$

(2) For $Z \in \Gamma TM$,

$$\begin{aligned}
g(\nabla(fk), Z) = Z(fk) &= fZ(k) + kZ(f) \\
&= fg(\nabla k, Z) + kg(\nabla f, Z) \\
&= g(f\nabla k + k\nabla f, Z).
\end{aligned}$$

Thus

$$\nabla(fk) = f\nabla k + k\nabla f.$$

(3) By (2),

$$\begin{aligned}
h_{fk}(X) &= \nabla_X \nabla(fk) \\
&= \nabla_X (f\nabla k + k\nabla f) \\
&= X(f)\nabla k + f\nabla_X \nabla k + X(k)\nabla f + k\nabla_X \nabla f \\
&= fh_k(X) + kh_f(X) + g(\nabla f, X)\nabla k + g(\nabla k, X)\nabla f.
\end{aligned}$$

(4) Obvious from (3).

(5) Let $\{X_1, \ldots, X_n\}$ be an orthonormal local frame. Then by (4),

$$\begin{aligned}\Delta(fk) &= -\operatorname{trace} h_{fk}\\ &= -\sum_{i=1}^{n} g(X_i, X_i) H_{fk}(X_i, X_i)\\ &= -f\sum_{i=1}^{n} g(X_i, X_i) H_k(X_i, X_i) - k\sum_{i=1}^{n} g(X_i, X_i) H_f(X_i, X_i)\\ &\quad -\sum_{i=1}^{n} g(X_i, X_i) g(\nabla f, X_i) g(\nabla k, X_i) - \sum_{i=1}^{n} g(X_i, X_i) g(\nabla k, X_i) g(\nabla f, X_i)\\ &= f\Delta k + k\Delta f - 2g(\nabla f, \nabla k).\end{aligned}$$

□

2.5 The Semi-Riemannian Divergence Theorem

Throughout this section, let (M, g) denote a semi-Riemannian manifold of dimension n and index ν with boundary ∂M (possibly $\partial M = \emptyset$).

The main difficulty in stating a divergence theorem for a semi-Riemannian manifold with boundary is that the restriction of the metric tensor to the boundary may fail to be nondegenerate at certain points of the boundary. Consequently there exists no unit outward normal to the boundary at such points, and hence, there is no canonical induced volume form on the boundary. In this section, we overcome these difficulties under some restrictions on the boundary and state a semi-Riemannian divergence theorem. First, we investigate the relation between the divergence of a vector field on (M, g) and the semi-Riemannian volume form of (M, g).

Definition 2.5.1 Let (M, g) be an oriented semi-Riemannian manifold with boundary ∂M (possibly $\partial M = \emptyset$). An exterior n-form μ on M is called a *semi-Riemannian volume form on* (M, g) if $\mu(X_1, \dots, X_n) = 1$ for every positively oriented orthonormal local frame.

Proposition 2.5.1 *Let (M, g) be an oriented semi-Riemannian manifold with boundary ∂M (possibly $\partial M = \emptyset$). Then there is a unique semi-Riemannian volume form μ on (M, g).*

Proof. Let $\{X_1, \dots, X_n\}$ be a positively oriented orthonormal local frame on an open set $U \subseteq M$ and let $\{\mu_1, \dots, \mu_n\}$ be the local co-frame on U dual to $\{X_1, \dots, X_n\}$. Then it can be shown by using elementary linear algebra that the exterior n-form μ_U defined by

$$\mu_U = \mu_1 \wedge \ldots \wedge \mu_n$$

is a semi-Riemannian volume form for (M, g) defined on U. Note that, if μ'_V is a semi-Riemannian volume form defined on some open set $V \subseteq M$ then $\mu'_V = \mu_U$ on $U \cap V$. Indeed, since the dimension of the space of alternating n-multilinear maps on an n-dimensional real vector space is 1, there exists a function h on $U \cap V$ such that $\mu'_V = h\mu_U$ on $U \cap V$, provided that $U \cap V \neq \emptyset$. Hence, since

$$\mu'_V(Y_1, \ldots, Y_n) = \mu_U(Y_1, \ldots, Y_n) = 1$$

for every positively oriented orthonormal local frame $\{Y_1, \ldots, Y_n\}$ on $U \cap V$, it follows that $h = 1$ on $U \cap V$. Thus, locally defined semi-Riemannian volume forms μ_U give a rise of a global semi-Riemannian volume form μ. The uniqueness of μ also follows from the argument above on μ'_V and μ_U. □

Let (M, g) be a semi-Riemannian manifold of index ν and $\{X_1, \ldots, X_\nu, X_{\nu+1}, \ldots, X_n\}$ be an orthonormal local frame with

$$g(X_i, X_i) = \begin{cases} -1 & \text{for} \quad 1 \leq i \leq \nu \\ 1 & \text{for} \quad \nu + 1 \leq i \leq n \end{cases}.$$

Then note that, if $\{\mu_1, \ldots, \mu_n\}$ is the local co-frame dual to $\{X_1, \ldots, X_n\}$ and if $\{\omega_1, \ldots, \omega_n\}$ is the local co-frame defined by $\omega_i = g^\flat(X_i) = g(X_i, \cdot)$ for $i = 1, \ldots, n$, where $g^\flat : \Gamma TM \to \Gamma TM^*$ is the musical isomorphism, then

$$\omega_i = -\mu_i \quad \text{for} \quad 1 \leq i \leq \nu$$

and

$$\omega_i = \mu_i \quad \text{for} \quad \nu + 1 \leq i \leq n.$$

Proposition 2.5.2 *Let (M, g) be an oriented semi-Riemannian manifold with boundary ∂M (possibly $\partial M = \emptyset$) and let μ be the semi-Riemannian volume form of (M, g). If Z is a vector field on M then*

$$\mathcal{L}_Z \mu = (\operatorname{div} Z)\mu,$$

where $\mathcal{L}$ is the Lie derivative on M.

Proof. It suffices to show that $\mathcal{L}_Z\mu = (\operatorname{div} Z)\mu$ at each $p \in M$. Now let $\{X_1, \ldots, X_\nu, X_{\nu+1}, \ldots, X_n\}$ be a positively oriented adapted moving frame near p with

$$g(X_i, X_i) = \begin{cases} -1 & \text{for} \quad 1 \leq i \leq \nu \\ 1 & \text{for} \quad \nu + 1 \leq i \leq n \end{cases}$$

and let $\{\mu_1, \ldots, \mu_n\}$ and $\{\omega_1, \ldots, \omega_n\}$ be co-frames defined as in the above. Then since $\mu = \mu_1 \wedge \ldots \wedge \mu_n$ near p,

$$\begin{aligned}
(\mathcal{L}_Z\mu)(X_1, \ldots, X_n)(p) &= \left[Z\mu(X_1, \ldots, X_n) - \sum_{i=1}^{n} \mu(X_1, \ldots, \mathcal{L}_Z X_i, \ldots, X_n) \right](p) \\
&= \left[-\sum_{i=1}^{n} \mu(X_1, \ldots, \nabla_Z X_i - \nabla_{X_i} Z, \ldots, X_n) \right](p) \\
&= \left[\sum_{i=1}^{n} \mu_1 \wedge \ldots \wedge \mu_n(X_1, \ldots, \nabla_{X_i} Z, \ldots, X_n) \right](p) \\
&= \left[\sum_{i=1}^{n} \mu_i(\nabla_{X_i} Z) \right](p) \\
&= \left[\sum_{i=1}^{\nu} (-\omega_i(\nabla_{X_i} Z)) + \sum_{i=\nu+1}^{n} \omega_i(\nabla_{X_i} Z) \right](p) \\
&= \left[\sum_{i=1}^{n} g(X_i, X_i) g(\nabla_{X_i} Z, X_i) \right](p) \\
&= (\text{div } Z)(p)\mu(X_1, \ldots, X_n)(p).
\end{aligned}$$

Hence,

$$(\mathcal{L}_Z\mu)(x_1, \ldots, x_n) = (\text{div } Z)(p)\mu(x_1, \ldots, x_n)$$

for every positively oriented orthonormal $\{x_1, \ldots, x_n\}$ in T_pM. Thus, it follows that

$$\mathcal{L}_Z\mu = (\text{div} Z)\mu.$$

□

Theorem 2.5.1 (Gauss Theorem) *Let (M, g) be an oriented semi-Riemannian manifold with boundary ∂M (possibly $\partial M = \emptyset$) and semi-Riemannian volume form μ. If Z is a vector field on M with compact support then*

$$\int_M (\text{div } Z)\mu = \int_{\partial M} i_Z\mu,$$

where i is the interior product.

Proof. Since $\mathcal{L}_Z = di_Z + i_Z d$, it follows that

$$(\text{div } Z)\mu = \mathcal{L}_Z\mu = di_Z\mu + i_Z d\mu = di_Z\mu.$$

Thus, by the Stoke's theorem,

$$\int_M (\text{div } Z)\mu = \int_M di_Z\mu = \int_{\partial M} i_Z\mu.$$

□

Let (M, g) be a semi-Riemannian manifold of index ν with boundary ∂M. Then the restriction $g_{\partial M}$ of g to the manifold ∂M is also a symmetric $(0, 2)$-tensor field on ∂M. But $g_{\partial M}$ may not be a metric tensor on the connected components of ∂M because it may fail to be nondegenerate at some points of the connected components of ∂M. The points of ∂M where $g_{\partial M}$ fails to be nondegenerate can also be characterized by the causal characters of the orthogonal vectors to ∂M. Note that, if $p \in \partial M$ then $T_p\partial M$ is a hyperspace in the inner product space (T_pM, g_p) with orthogonal space $(T_p\partial M)^\perp$ of dimension 1 by Proposition 1.1.1. Thus by Proposition 1.1.2, $(T_p\partial M, (g_{\partial M})_p)$ is a degenerate hyperspace in (T_pM, g_p) if and only if $((T_p\partial M)^\perp, g_{p(T_p\partial M)^\perp})$ is a degenerate subspace in (T_pM, g_p) if and only if nonzero vectors orthogonal to $T_p\partial M$ are null vectors (contained in $T_p\partial M$). Hence, $(T_p\partial M, (g_{\partial M})_p)$ is nondegenerate if and only if nonzero vectors orthogonal to $T_p\partial M$ are nonnull. In fact, if $(T_p\partial M, (g_{\partial M})_p)$ is nondegenerate then, nonzero vectors orthogonal to $T_p\partial M$ are spacelike if and only if $(g_{\partial M})_p$ is nondegenerate of index ν, and nonzero vectors orthogonal to $T_p\partial M$ are timelike if and only if $(g_{\partial M})_p$ is nondegenerate of index $\nu - 1$.

Definition 2.5.2 Let (M, g) be a semi-Riemannian manifold of index ν with boundary ∂M. Define the subsets ∂M_+ and ∂M_- of ∂M to be the sets of points in ∂M where the nonzero vectors orthogonal to ∂M are spacelike and timelike, respectively. Also define the subset ∂M_0 of ∂M to be the set of points in ∂M where the nonzero orthogonal vectors to ∂M are null.

Remark 2.5.1 Note that $\partial M = \partial M_+ \cup \partial M_- \cup \partial M_0$ and ∂M_+, ∂M_- and ∂M_0 are pairwise disjoint.

Lemma 2.5.1 *Let (M, g) be a semi-Riemannian manifold of index ν with boundary ∂M. Then ∂M_+ and ∂M_- are open subsets of ∂M.*

Proof. Since ∂M_+ and ∂M_- are the sets of points of ∂M where $g_{\partial M}$ is nondegenerate of index ν and $\nu - 1$, respectively, it follows from Lemma 2.1.1 that ∂M_+ and ∂M_- are open subsets of ∂M. □

Remark 2.5.2 Note that, it follows from Lemma 2.5.1 that ∂M_+ and ∂M_- are open submanifolds of ∂M. Hence $\partial M' = \partial M_+ \cup \partial M_-$ is also an open submanifold of ∂M which we call the *nondegenerate boundary* of (M, g).

Let (M, g) be an oriented semi-Riemannian manifold with boundary ∂M. Let μ be the semi-Riemannian volume form on (M, g) and let N be the unit outward normal vector field to ∂M defined on $\partial M'$. Then note that, the exterior $(n-1)$-form $\mu_{\partial M'}$ defined on $\partial M'$ by

$$\mu_{\partial M'} = i_N \mu$$

is the induced semi-Riemannian volume forms on semi-Riemannian manifolds $(\partial M_+, g_{\partial M_+})$ of index ν and $(\partial M_-, g_{\partial M_-})$ of index $\nu - 1$ when restricted to ∂M_+ and ∂M_-, respectively. By using the notation above, we state the semi-Riemannian divergence theorem as follows.

Theorem 2.5.2 [Ü] **(The Semi-Riemannian Divergence Theorem)** *Let (M, g) be an oriented semi-Riemannian manifold with boundary ∂M (possibly $\partial M = \emptyset$) and semi-Riemannian volume form μ. Let Z be a vector field on M with compact support. If ∂M_0 has measure zero in ∂M (cf. [BJ, p. 56]) then*

$$\int_M (\text{div } Z)\mu = \int_{\partial M_+} g(Z, N)\mu_{\partial M'} - \int_{\partial M_-} g(Z, N)\mu_{\partial M'}.$$

Proof. Since ∂M_0 has measure zero in ∂M, by Theorem 2.5.1,

$$\int_M (\text{div } Z)\mu = \int_{\partial M} i_Z\mu = \int_{\partial M'} i_Z\mu = \int_{\partial M_+} i_Z\mu + \int_{\partial M_-} i_Z\mu.$$

Now let $p \in \partial M'$ and let $\{N(p), e_1, \ldots, e_{n-1}\}$ be an orthonormal basis for (T_pM, g_p). Then at p, since

$$Z = g(N, N)g(N, Z)N + \sum_{i=1}^{n-1} g(e_i, e_i)g(Z, e_i)e_i,$$

$$\begin{aligned}(i_Z\mu)(e_1,\ldots,e_{n-1}) &= \mu(Z,e_1,\ldots,e_{n-1})\\ &= \mu\,(g(N,N)g(N,Z)N\\ &\quad +\sum_{i=1}^{n-1} g(e_i,e_i)g(Z,e_i)e_i,e_1,\ldots,e_{n-1})\\ &= \mu(g(N,N)g(N,Z)N,e_1,\ldots,e_{n-1})\\ &= g(N,N)g(N,Z)\mu(N,e_1,\ldots,e_{n-1})\\ &= g(N,N)g(N,Z)(i_N\mu)(e_1,\ldots,e_{n-1})\\ &= g(N,N)g(N,Z)\mu_{\partial M'}(e_1,\ldots,e_{n-1}).\end{aligned}$$

Hence

$$i_Z\mu = \begin{cases} g(Z,N)\mu_{\partial M'} & \text{on } \partial M_+ \\ -g(Z,N)\mu_{\partial M'} & \text{on } \partial M_-. \end{cases}.$$

Thus it follows that

$$\int_M (\text{div } Z)\mu = \int_{\partial M_+} g(Z,N)\mu_{\partial M'} - \int_{\partial M_-} g(Z,N)\mu_{\partial M'}.$$

□

The assumption of vanishing measure of ∂M_0 in ∂M may be replaced by another condition in the semi-Riemannian divergence theorem.

Theorem 2.5.3 [Ü] *Let (M,g) be an oriented semi-Riemannian manifold with boundary ∂M (possibly $\partial M = \emptyset$) and semi-Riemannian volume form μ. Let Z be a vector field on M with compact support. If Z is tangent to ∂M at the points of ∂M_0 then*

$$\int_M (\text{div } Z)\mu = \int_{\partial M_+} g(Z,N)\mu_{\partial M'} - \int_{\partial M_-} g(Z,N)\mu_{\partial M'}.$$

Proof. Since Z is tangent to ∂M at the points of ∂M_0, $i_Z\mu = 0$ at the points of ∂M_0. Then by Theorem 2.5.1,

$$\int_M (\text{div } Z)\mu = \int_{\partial M} i_Z\mu = \int_{\partial M'} i_Z\mu = \int_{\partial M_+} i_Z\mu + \int_{\partial M_-} i_Z\mu.$$

Then as in Theorem 2.5.2, we obtain

$$\int_M (\text{div } Z)\mu = \int_{\partial M_+} g(Z,N)\mu_{\partial M'} - \int_{\partial M_-} g(Z,N)\mu_{\partial M'}.$$

□

The Theorems 2.5.2 and 2.5.3 fail to hold if we remove the assumptions on ∂M_0. For example, let $M = S^1 \times [-1,1]$ with metric tensor

$$g = \frac{1}{2}[d\theta \otimes dt + dt \otimes d\theta] + (1-t^2)d\theta \otimes d\theta,$$

where θ is the polar coordinate on S^1 and t is the usual coordinate on $[-1,1]$. Then it is easy to see that (M,g) is a Lorentzian manifold with boundary

$$\partial M = (S^1 \times \{1\}) \cup (S^1 \times \{-1\}) = \partial M_0.$$

Now orient M by $[\frac{\partial}{\partial\theta}, \frac{\partial}{\partial t}]$. Then with a simple computation, it can be shown that the semi-Riemannian volume form of (M,g) is

$$\mu = \frac{1}{2} d\theta \wedge dt.$$

Now consider $Z = t\frac{\partial}{\partial t}$ on M. It can be easily shown that div $Z = 1$ and hence

$$\int_M (\text{div } Z)\mu = 2\pi.$$

But Theorems 2.5.2 and 2.5.3 predict

$$\int_M (\text{div } Z)\mu = 0$$

as a wrong answer.

Now we make an application of the semi-Riemannian divergence theorem to subharmonic functions.

Theorem 2.5.4 *Let (M,g) be an oriented compact semi-Riemannian manifold. Then every subharmonic function on (M,g) is a harmonic function on (M,g).*

Proof. Let $f : (M,g) \to \mathbf{R}$ be a subharmonic function on (M,g). Then, since $\partial M = \emptyset$, by Theorem 2.5.2,

$$\int_M -(\Delta f)\mu = \int_M (\text{div } \nabla f)\mu = 0,$$

where μ is the semi-Riemannian volume form of (M,g). Since $-\Delta f \geq 0$, it follows that $\Delta f = 0$, that is, f is harmonic on (M,g). □

It is worth noting here that if (M, g) is a Riemannian manifold then we have the following stronger version of Theorem 2.5.4.

Theorem 2.5.5 *Let (M, g) be a connected, compact, oriented Riemannian manifold. Then every subharmonic function on (M, g) is constant.*

Proof. Let $f : (M, g) \to \mathbf{R}$ be a subharmonic function on (M, g). First note that by Proposition 2.4.2–(5),

$$\frac{1}{2}\Delta f^2 = f\Delta f - g(\nabla f, \nabla f).$$

Also since $\Delta f = 0$ by Theorem 2.5.4, we have

$$\frac{1}{2}\Delta f^2 = -g(\nabla f, \nabla f).$$

On the other hand, since $\partial M = \emptyset$,

$$\int_M (\Delta f^2)\mu = -\int_M (\text{div } \nabla f^2)\mu = 0$$

by Theorem 2.5.2, where μ is the Riemannian volume form of (M, g).

Thus

$$\int_M g(\nabla f, \nabla f)\mu = 0$$

and it follows that $\nabla f = 0$ since g is a positive definite metric tensor. Hence f is constant on (M, g). □

Note that Theorem 2.5.5 is not valid for compact semi-Riemannian manifolds in general. For example, let $M = S^1 \times S^1$ with metric tensor

$$g = -d\theta^1 \otimes d\theta^1 + d\theta^2 \otimes d\theta^2,$$

where θ^1 and θ^2 are polar coordinates on the first and second factors of M, respectively. Then it is easy to show that the function

$$f(\theta^1, \theta^2) = \sin(\theta^1 + \theta^2)$$

is a harmonic function on (M, g). Yet clearly f is not constant on (M, g). For a discussion of harmonic functions on semi-Riemannian manifolds, see [K6].

Also later in Chapter 3, we will generalize the semi-Riemannian divergence theorem to vector fields along a map. (See Theorem 3.3.1).

Finally we make an application of Theorems 2.5.2 and 2.5.3 in 2-dimensional Minkowski space $(\mathbf{R}^2, \cdot)$, where $\cdot = dx^1 \otimes dx^1 - dx^2 \otimes dx^2$.

Let C be a curve in $\mathbf{R}^2$ parametrized by $\gamma = (\gamma^1, \gamma^2) : I \to \mathbf{R}^2$, where I is either of the intervals $[a, b]$, $[a, b)$, $(a, b]$ or (a, b). The *line element* of C in $(\mathbf{R}^2, \cdot)$ is defined by

$$ds = \left| \left(\frac{d\gamma^1}{dt}\right)^2 - \left(\frac{d\gamma^2}{dt}\right)^2 \right|^{1/2} dt.$$

Let f be a function defined on C. Then the *line integral* of f over C in $(\mathbf{R}^2, \cdot)$ is defined by

$$\int_C f ds = \int_a^b f(\gamma^1, \gamma^2) \left| \left(\frac{d\gamma^1}{dt}\right)^2 - \left(\frac{d\gamma^2}{dt}\right)^2 \right|^{1/2} dt,$$

whenever the integral (possibly improper) on the right hand side exists.

Now let D be a region in $\mathbf{R}^2$ bounded by a curve C parametrized by $\gamma = (\gamma^1, \gamma^2) : I \to \mathbf{R}^2$ which is regular and one to one. Note that, if we restrict $\cdot$ to the closure $cl(D)$ of D then $(cl(D), \cdot)$ is a 2-dimensional Lorentzian manifold with boundary $\partial(cl(D)) = C$. Set

$$C_+ = \partial(cl(D))_+$$
$$C_- = \partial(cl(D))_-$$
$$C_0 = \partial(cl(D))_0.$$

Note that C_+ (resp., C_-) is traced out by the portions of γ defined on disjoint open subintervals of I where $\dot{\gamma}$ is timelike (resp., spacelike). Hence, if f is a function defined on C, define

$$\int_{C_+} f ds \qquad (\text{resp., } \int_{C_-} f ds)$$

to be the (finite) sum of line integrals of f over the portions of γ defined on disjoint open subintervals of I where $\dot{\gamma}$ is timelike (resp., spacelike).

Theorem 2.5.6 [Ü] *Let D be a region in 2-dimensional Minkowski space $(\mathbf{R}^2, \cdot)$ bounded by a positively oriented curve C parametrized by $\gamma = (\gamma^1, \gamma^2) : I \to \mathbf{R}^2$ which is regular and one to one. (Positively oriented means the region D is on your left as you traverse the curve in the positive direction). Let Z be a vector field on $cl(D)$. If C_0 has measure zero in C or Z is tangent to C at the points of C_0, then*

$$\int\int_D (\text{div } Z) dx^1 dx^2 = \int_{C_+} (Z \cdot N) ds - \int_{C_-} (Z \cdot N) ds,$$

where N is the unit outward normal vector field to C defined on $C' = C_+ \cup C_-$.

Proof. Orient $\mathbf{R}^2$ by $[\frac{\partial}{\partial x^1}, \frac{\partial}{\partial x^2}]$. Then note that the semi-Riemannian volume form μ of $(\mathbf{R}^2, \cdot)$ is given by

$$\mu = dx^1 \wedge dx^2.$$

Now since

$$\int_{cl(D)} (\text{div} Z)\mu = \int_D (\text{div} Z)\mu = \int\int_D (\text{div} Z) dx^1 dx^2,$$

it follows from Theorems 2.5.2 and 2.5.3 that

$$\int\int_D (\text{div} Z) dx^1 dx^2 = \int_{C_+} (Z \cdot N)\mu_{C'} - \int_{C_-} (Z \cdot N)\mu_{C'}.$$

Now note that

$$N \circ \gamma = \begin{cases} \dfrac{\left(\frac{d\gamma^2}{dt}, \frac{d\gamma^1}{dt}\right)}{\left|\left(\frac{d\gamma^1}{dt}\right)^2 - \left(\frac{d\gamma^2}{dt}\right)^2\right|^{1/2}} & \text{on } C_+ \\ -\dfrac{\left(\frac{d\gamma^2}{dt}, \frac{d\gamma^1}{dt}\right)}{\left|\left(\frac{d\gamma^1}{dt}\right)^2 - \left(\frac{d\gamma^2}{dt}\right)^2\right|^{1/2}} & \text{on } C_- \end{cases}.$$

Hence with an easy computation it can be shown that

$$\begin{aligned} \gamma^* \mu_{C_+} &= \frac{1}{\left|\left(\frac{d\gamma^1}{dt}\right)^2 - \left(\frac{d\gamma^2}{dt}\right)^2\right|^{1/2}} \left(\left(\frac{d\gamma^2}{dt}\right)^2 - \left(\frac{d\gamma^1}{dt}\right)^2\right) dt \\ &= \left|\left(\frac{d\gamma^1}{dt}\right)^2 - \left(\frac{d\gamma^2}{dt}\right)^2\right|^{1/2} dt = ds \end{aligned}$$

and

$$\begin{aligned}
\gamma^*\mu_{C_-} &= \frac{1}{\left|\left(\frac{d\gamma^1}{dt}\right)^2 - \left(\frac{d\gamma^2}{dt}\right)^2\right|^{1/2}} \left(\left(\frac{d\gamma^1}{dt}\right)^2 - \left(\frac{d\gamma^2}{dt}\right)^2\right) dt \\
&= \left|\left(\frac{d\gamma^1}{dt}\right)^2 - \left(\frac{d\gamma^2}{dt}\right)^2\right|^{1/2} dt = ds.
\end{aligned}$$

Thus it follows that

$$\int\int_D (\text{div} Z) dx^1 dx^2 = \int_{C_+} (Z \cdot N) ds - \int_{C_-} (Z \cdot N) ds.$$

□

Chapter 3

Second Fundamental Form of a Map

In this chapter, we introduce the second fundamental form of a map between semi-Riemannian manifolds. The second fundamental form of a map is used largely in the analysis of certain properties of maps between manifolds in the literature. However, in this chapter, we study the second fundamental form of a map only as much as we need to carry out our work on semi-Riemannian maps. We devote the first section to the definition and basic properties of the second fundamental form of a map between semi-Riemannian manifolds. In section 2, we define affine and harmonic maps between semi-Riemannian manifolds and derive certain properties of affine maps. Then in section 3, we define the divergence of a vector field along a map between semi-Riemannian manifolds and generalize the semi-Riemannian divergence theorem to such vector fields. Finally in section 4, we derive the Bochner identity of a map between semi-Riemannian manifolds which will be used in showing a Riemannian map to be affine under certain curvature conditions. Throughout this chapter, and in all following chapters, let M_1, M_2 and M be manifolds with dimensions n_1, n_2 and n, respectively. Also let (M_1, g_1), (M_2, g_2) and (M, g) be semi-Riemannian manifolds of indices ν_1, ν_2 and ν with Levi-Civita connections $\overset{1}{\nabla}$, $\overset{2}{\nabla}$ and ∇, respectively, unless otherwise stated.

3.1 The Second Fundamental Form

In this section, we make the main definitions which will be used throughout this book. Let $f : M_1 \to M_2$ be a map. Recall that, if $X \in \Gamma TM_1$ then $f_* X \in \Gamma_f TM_2$. (See Remark 2.2.1). Hence we can consider f_* as a map

$$f_* : \Gamma TM_1 \to \Gamma_f TM_2.$$

It is easy to show that f_* is linear in its argument. For example, let $h \in C^\infty(M_1)$ and $X \in \Gamma TM_1$. Then if we set $\pi_1 : TM_1 \to M_1$ to be the canonical projection,

$$\begin{aligned} f_*(hX) &= f_{*\pi_1(hX)} h(\pi_1(X))X(\pi_1(X)) \\ &= h(\pi_1(X)) f_{*\pi_1(X)}(X(\pi_1(X))) \\ &= hf_*X \end{aligned}$$

since $f_{*\pi_1(X)} : T_{\pi_1(X)}M_1 \to T_{f(\pi_1(X))}M_2$ is a linear map.

Definition 3.1.1 Let $f : (M_1, g_1) \to (M_2, g_2)$ be a map. The *second fundamental form of* f is the map

$$\nabla f_* : \Gamma TM_1 \times \Gamma TM_1 \to \Gamma_f TM_2$$

defined by

$$(\nabla f_*)(X, Y) = \overset{2}{\nabla}_X f_*Y - f_*(\overset{1}{\nabla}_X Y),$$

where $\overset{2}{\nabla}$ also denotes the pullback of $\overset{2}{\nabla}$ along f.

It is easy to check that ∇f_* is linear in its arguments. For example, let $h \in C^\infty(M_1)$ and $X, Y \in \Gamma TM_1$. Then

$$\begin{aligned} (\nabla f_*)(X, hY) &= \overset{2}{\nabla}_X f_*(hY) - f_*(\overset{1}{\nabla}_X (hY)) \\ &= \overset{2}{\nabla}_X (hf_*Y) - f_*(X(h)Y + h\overset{1}{\nabla}_X Y) \\ &= X(h)f_*Y + h\overset{2}{\nabla}_X f_*Y \\ &\quad -X(h)f_*Y - hf_*(\overset{1}{\nabla}_X Y) \\ &= h((\overset{2}{\nabla}_X f_*Y) - f_*(\overset{1}{\nabla}_X Y)) \\ &= h(\nabla f_*)(X, Y). \end{aligned}$$

Note that, it can also be shown, as in Proposition 2.1.1, that the linearity of ∇f_* in its arguments implies that the value of $(\nabla f_*)(X, Y)$ at $p_1 \in M_1$ only depends on the values of X and Y at $p_1 \in M_1$. Thus one can define $(\nabla f_*)(x, y)$ by

$$(\nabla f_*)(x, y) = (\nabla f_*)(X, Y)(p_1)$$

where $X, Y \in \Gamma TM_1$ with $X(p_1) = x$ and $Y(p_1) = y$.

Proposition 3.1.1 *Let $f : (M_1, g_1) \to (M_2, g_2)$ be a map. Then*

$$(\nabla f_*)(X,Y) = (\nabla f_*)(Y,X)$$

for all X, $Y \in \Gamma TM_1$, that is, ∇f_* is symmetric.

Proof. By Theorem 2.2.3–(b),

$$\begin{aligned}(\nabla f_*)(X,Y) &= \overset{2}{\nabla}_X f_*Y - f_*(\overset{1}{\nabla}_X Y)\\ &= \overset{2}{\nabla}_Y f_*X + f_*[X,Y] - f_*(\overset{1}{\nabla}_Y X) - f_*[X,Y]\\ &= \overset{2}{\nabla}_Y f_*X - f_*(\overset{1}{\nabla}_Y X)\\ &= (\nabla f_*)(Y,X).\end{aligned}$$

□

Definition 3.1.2 Let $f : (M_1, g_1) \to (M_2, g_2)$ be a map. Then for each fixed $X \in \Gamma TM_1$, the linear map

$$\nabla_X f_* : \Gamma TM_1 \to \Gamma_f TM_2$$

is defined by

$$(\nabla_X f_*)(Y) = (\nabla f_*)(X,Y).$$

Observe that the linearity of $\nabla_X f_*$ follows from the linearity of ∇f_* in its arguments. Also, since the value of $(\nabla f_*)(X,Y)$ at $p_1 \in M_1$ only depends on the value of X and Y at $p_1 \in M_1$, for each fixed $x \in T_{p_1}M_1$, we can define a linear transformation

$$\nabla_x f_* : T_{p_1}M_1 \to T_{f(p_1)}M_2$$

by

$$(\nabla_x f_*)(y) = ((\nabla_X f_*)(Y))(p_1) = (\nabla f_*)(X,Y)(p_1),$$

where $X, Y \in \Gamma TM_1$ with $X(p_1) = x$ and $Y(p_1) = y$.

Now let $f : (M_1, g_1) \to (M_2, g_2)$ be a map. Then note that, for each $p_1 \in M_1$,

$$f_{*p_1} : (T_{p_1}M_1, g_{1p_1}) \to (T_{f(p_1)}M_1, g_{2f(p_1)})$$

is a linear transformation between inner product spaces $(T_{p_1}M_1, g_{1p_1})$ and $(T_{f(p_1)}M_1, g_{2f(p_1)})$.

Definition 3.1.3 Let $f : (M_1, g_1) \to (M_2, g_2)$ be a map. Then the *square norm of* f_* is the map

$$\|f_*\|^2 : M_1 \to \mathbf{R}$$

defined by

$$\|f_*\|^2(p_1) = \|f_{*p_1}\|^2.$$

(See Definition 1.2.3).

Note that, since

$$\|f_*\|^2 = \sum_{i=1}^{n_1} g_1(X_i, X_i) g_2(f_* X_i, f_* X_i),$$

where $\{X_1, \ldots, X_{n_1}\}$ is an orthonormal local frame for TM_1, $\|f\|^2$ is a (smooth) map on M_1. Note here that $\|f_*\|^2$ can also be considered to be the trace of $f^* g_2$ with respect to g_1 by the expression above.

Now we compute the gradient of $\|f_*\|^2$ on (M_1, g_1) which will be used in the computation of the Bochner identity and showing the fact that $\|f_*\|^2$ is locally constant for affine maps.

Note that, if $f : (M_1, g_1) \to (M_2, g_2)$ is a map then, for each $x \in T_{p_1} M_1$,

$$\nabla_x f_* : (T_{p_1} M_1, g_{1p_1}) \to (T_{f(p_1)} M_2, g_{2f(p_1)})$$

is a linear transformation between inner product spaces $(T_{p_1} M_1, g_{1p_1})$ and $(T_{f(p_1)} M_2, g_{2f(p_1)})$. Let ${}^*(\nabla_x f_*)$ denote the adjoint of $\nabla_x f_*$. (See Definition 1.2.2).

Lemma 3.1.1 *Let* $f : (M_1, g_1) \to (M_2, g_2)$ *be a map. Then*

$$(\overset{1}{\nabla} \|f_*\|^2)(p_1) = 2 \sum_{i=1}^{n_1} g_1(x_i, x_i) \left({}^*(\nabla_{x_i} f_*) \circ f_{*p_1} \right) x_i,$$

where $\{x_1, \ldots, x_{n_1}\}$ *is an orthonormal basis for* $(T_{p_1} M_1, g_{1p_1})$ *and* $\overset{1}{\nabla} \|f_*\|^2$ *denotes the gradient of* $\|f_*\|^2$ *on* (M_1, g_1).

Proof. Let $\{x_1, \ldots, x_{n_1}\}$ be an orthonormal basis for $(T_{p_1} M_1,\ g_{1p_1})$ and let $\{X_1, \ldots, X_{n_1}\}$ be the extension of $\{x_1, \ldots, x_{n_1}\}$ to an adapted moving frame near $p_1 \in M_1$. Then for $X \in \Gamma TM_1$,

$$\begin{aligned}
g_1((\overset{1}{\nabla}\|f_*\|^2)(p_1), X(p_1)) &= X(p_1)\|f_*\|^2 \\
&= \left[\sum_{i=1}^{n_1} g_1(X_i, X_i) X(g_2(f_*X_i, f_*X_i))\right](p_1) \\
&= 2\left[\sum_{i=1}^{n_1} g_1(X_i, X_i) g_2(\overset{2}{\nabla}_X\, f_*X_i, f_*X_i)\right](p_1) \\
&= 2\left[\sum_{i=1}^{n_1} g_1(X_i, X_i) g_2((\nabla f_*)(X, X_i), f_*X_i)\right](p_1) \\
&= 2\left[\sum_{i=1}^{n_1} g_1(X_i, X_i) g_2((\nabla f_*)(X_i, X), f_*X_i)\right](p_1) \\
&= 2\left[\sum_{i=1}^{n_1} g_1(X_i, X_i) g_2((\nabla_{X_i} f_*)X, f_*X_i)\right](p_1) \\
&= 2\sum_{i=1}^{n_1} g_1(x_i, x_i) g_2((\nabla_{x_i} f_*)X(p_1), f_{*p_1}x_i) \\
&= 2\sum_{i=1}^{n_1} g_1(x_i, x_i) g_1(X(p_1), ({}^*(\nabla_{x_i} f_*) \circ f_{*p_1})x_i) \\
&= g_1(2\sum_{i=1}^{n_1} g_1(x_i, x_i) g_1({}^*(\nabla_{x_i} f_*) \circ f_{*p_1})x_i, X(p_1)).
\end{aligned}$$

Thus

$$(\overset{1}{\nabla}\|f_*\|^2)(p_1) = 2\sum_{i=1}^{n_1} g_1(x_i, x_i)({}^*(\nabla_{x_i} f_*) \circ f_{*p_1})x_i.$$

□

We also have the following corollary to the Lemma above.

Corollary 3.1.1 *Let $f : (M, g) \to \mathbf{R}$ be a map. Then*

$$\nabla g(\nabla f, \nabla f) = 2\nabla_{\nabla f}\nabla f,$$

where ∇f and $\nabla g(\nabla f, \nabla f)$ denote the gradients of f and $g(\nabla f, \nabla f)$ on (M, g), respectively.

Proof. If we set $(M_1, g_1) = (M, g)$ and $(M_2, g_2) = (\mathbf{R}, dt \otimes dt)$ in Lemma 3.1.1, by observing that, for $X \in \Gamma TM$,

$$f_* X = g(\nabla f, X)\frac{d}{dt} \circ f,$$

it can be shown that

$$\|f_*\|^2(p) = g(\nabla f, \nabla f)(p)$$

and

$$2\sum_{i=1}^{n_1} g_1(x_i, x_i)({}^*(\nabla_{x_i} f_*) \circ f_{*p_1})x_i = 2(\nabla_{\nabla f}\nabla f)(p),$$

where $p_1 = p \in M_1 = M$. Hence by Lemma 3.1.1, the claim follows. We leave these computations as an instructive exercise to the reader and compute $\nabla g(\nabla f, \nabla f)$ directly which is much shorter. Indeed, if $X \in \Gamma TM$ then,

$$\begin{aligned} g(\nabla g(\nabla f, \nabla f), X) &= Xg(\nabla f, \nabla f) \\ &= 2g(\nabla_X \nabla f, \nabla f) \\ &= 2g(h_f(X), \nabla f) \\ &= 2g(h_f(\nabla f), X) \\ &= 2g(\nabla_{\nabla f}\nabla f, X), \end{aligned}$$

where h_f is the Hessian tensor of f on (M, g). Thus,

$$\nabla g(\nabla f, \nabla f) = 2\nabla_{\nabla f}\nabla f.$$

□

3.2 Affine and Harmonic Maps

In this section, we define affine and harmonic maps between semi-Riemannian manifolds by using the second fundamental form. First we deal with affine maps.

Definition 3.2.1 A map $f : (M_1, g_1) \to (M_2, g_2)$ is called *affine* if it has vanishing second fundamental form, that is, $\nabla f_* = 0$.

Next we will discuss some conditions which characterize affine maps.

Let $f : (M_1, g_1) \longrightarrow (M_2, g_2)$ be a map and $\gamma : I \longrightarrow M_1$ be a curve, where $I \subseteq \mathbf{R}$ is an open interval. Then note that, $f \circ \gamma : I \longrightarrow M_2$ is also a curve and for each given vector field X along γ, we can define a vector field $f_* X$ along $f \circ \gamma$ by

$$(f_* X)(t) = f_{*\gamma(t)} X(t),$$

where $t \in I$. We say that f *preserves parallel translation* if for every curve $\gamma : I \longrightarrow M_1$ and for every parallel X along γ, $f_* X$ is parallel along $f \circ \gamma : I \longrightarrow M_2$.

As a notation, let $\overset{1}{\nabla}$ also denote the pullback of the Levi-Civita connection $\overset{1}{\nabla}$ of (M_1, g_1) along γ and let $\overset{2}{\nabla}$ also denote both pullbacks of the Levi-Civita connection $\overset{2}{\nabla}$ of (M_2, g_2) along f and $f \circ \gamma$.

Proposition 3.2.1 *Let $f : (M_1, g_1) \longrightarrow (M_2, g_2)$ be a map. Then, f is affine if and only if f preserves parallel translation.*

Proof. Let $\gamma : I \longrightarrow M_1$ be a curve, where $I \subseteq \mathbf{R}$ is an open interval, and let X be a parallel vector field along γ. Now let X' be a locally defined vector field on M_1 with $X = X' \circ \gamma$. Note that, then

$$f_* X = (f_* X') \circ \gamma$$

and hence,

$$\begin{aligned}
\overset{2}{\nabla}_{\frac{d}{dt}} f_* X &= \overset{2}{\nabla}_{\frac{d}{dt}} ((f_* X') \circ \gamma) \\
&= \overset{2}{\nabla}_{\gamma_* \frac{d}{dt}} f_* X' \\
&= \overset{2}{\nabla}_{\dot{\gamma}} f_* X' \\
&= (\nabla f_*)(\dot{\gamma}, X' \circ \gamma) + f_*(\overset{1}{\nabla}_{\dot{\gamma}} X') \\
&= (\nabla f_*)(\dot{\gamma}, X) + f_*(\overset{1}{\nabla}_{\frac{d}{dt}} (X' \circ \gamma)) \\
&= (\nabla f_*)(\dot{\gamma}, X) + f_*(\overset{1}{\nabla}_{\frac{d}{dt}} X) \\
&= (\nabla f_*)(\dot{\gamma}, X).
\end{aligned}$$

Thus, f is affine if and only if f preserves parallel translation. □

Another characterization of affine maps is given in terms of geodesics. A map $f : (M_1, g_1) \to (M_2, g_2)$ is called to send *geodesics to geodesics*, if $\gamma : I \to M_1$ is a geodesic of (M_1, g_1) then $f \circ \gamma : I \to M_2$ is also a geodesic of (M_2, g_2).

Proposition 3.2.2 *Let $f : (M_1, g_1) \to (M_2, g_2)$ be a map. Then, f is affine if and only if f sends geodesics to geodesics.*

Proof. Let $\gamma : I \to M_1$ be a geodesic of (M_1, g_1) and let $\alpha = f \circ \gamma$. Then since $\dot{\gamma}$ is a parallel vector field along γ and $\dot{\alpha} = f_* \dot{\gamma}$ is a vector field along α, it follows as in the above proposition by setting $X = \dot{\gamma}$ that

$$\overset{2}{\nabla}_{\frac{d}{dt}} \dot{\alpha} = (\nabla f_*)(\dot{\gamma}, \dot{\gamma}).$$

By the fact that ∇f_* is symmetric, it follows that f is affine if and only if f sends geodesics to geodesics. □

Proposition 3.2.3 *Let $f : (M_1, g_1) \to (M_2, g_2)$ be an affine map. Then*

(1) *$f \circ \exp_1 = \exp_2 \circ f_*$, where $\exp_1$ and $\exp_2$ are the exponential maps of (M_1, g_1) and (M_2, g_2), respectively.*

(2) *$f_* \overset{1}{R}(X, Y)Z = \overset{2}{R}(f_*X, f_*Y)f_*Z$, where $X, Y, Z \in \Gamma TM_1$ and, $\overset{1}{R}$ and $\overset{2}{R}$ are the curvature tensors of (M_1, g_1) and (M_2, g_2), respectively.*

Proof. (1) Let $v \in T_{p_1}M_1$ be in the domain $\overset{1}{\mathcal{D}}$ of the exponential map of (M_1, g_1) and assume that $f_*v \in T_{f(p_1)}M_2$ be in the domain $\overset{2}{\mathcal{D}}$ of the exponential map of (M_2, g_2). Now let $\gamma_v : I \to M_1$ be a geodesic of (M_1, g_1) with $\gamma_v(0) = p_1$ and $\dot{\gamma}_v(0) = v$. Then by Proposition 3.2.2, since $\alpha = f \circ \gamma_v : I \to M_2$ is a geodesic of (M_2, g_2) with $\dot{\alpha}(0) = f_{*p_1}(\dot{\gamma}_v(0)) = f_{*p_1}v$, we have

$$\begin{aligned} (f \circ \exp_1)(v) &= f(\exp_1(v)) = f(\gamma_v(1)) = (f \circ \gamma_v)(1) \\ &= \alpha(1) = \exp_2(\dot{\alpha}(0)) = \exp_2(f_{*p_1}v) \\ &= (\exp_2 \circ f_*)(v). \end{aligned}$$

Thus, $f \circ \exp_1 = \exp_2 \circ f_*$.

(2) Let $\overset{2}{R}$ also denote the curvature tensor of the pullback of the Levi-Civita connection of (M_2, g_2) along f, and as well, let $\overset{2}{\nabla}$ also denote the pullback of the Levi-Civita connection $\overset{2}{\nabla}$ of (M_2, g_2) along f. Then by Proposition 2.3.1,

$$\begin{aligned}
\overset{2}{R}(f_*X, f_*Y)f_*Z &= \overset{2}{R}(X,Y)f_*Z \\
&= \overset{2}{\nabla}_X\overset{2}{\nabla}_Y f_*Z - \overset{2}{\nabla}_Y\overset{2}{\nabla}_X f_*Z - \overset{2}{\nabla}_{[X,Y]} f_*Z \\
&= \overset{2}{\nabla}_X (f_*(\overset{1}{\nabla}_Y Z)) - \overset{2}{\nabla}_Y (f_*(\overset{1}{\nabla}_X Z)) - f_*(\overset{1}{\nabla}_{[X,Y]} Z) \\
&= f_*(\overset{1}{\nabla}_X\overset{1}{\nabla}_Y Z) - f_*(\overset{1}{\nabla}_Y\overset{1}{\nabla}_X Z) - f_*(\overset{1}{\nabla}_{[X,Y]} Z) \\
&= f_*(\overset{1}{\nabla}_X\overset{1}{\nabla}_Y Z - \overset{1}{\nabla}_Y\overset{1}{\nabla}_X Z - \overset{1}{\nabla}_{[X,Y]} Z) \\
&= f_* \overset{1}{R}(X,Y)Z. \qquad \square
\end{aligned}$$

Proposition 3.2.4 *Let $f : (M_1, g_1) \to (M_2, g_2)$ be an affine map, where M_1 is connected. Then,*

(1) *rank of f is constant on M_1,*

(2) *$\|f_*\|^2$ is constant on M_1.*

Proof. (1) Let $p_1 \in M_1$ with dim (ker f_{*p_1}) $= k$ and let $q_1 \in M_1$. We will show that, also dim (ker f_{*q_1}) $= k$. Hence it follows that $\text{rank}_{p_1} f = \text{rank}_{q_1} f$, that is, f is of constant rank. For this, let $\gamma : I \to M_1$ be a curve with $\gamma(t_1) = p_1$ and $\gamma(t_2) = q_1$. Now let $\{x_1, \ldots, x_k\}$ be a basis for ker f_{*p_1} and $\{X_1, \ldots, X_k\}$ be the parallel vector fields along γ with $X_i(t_1) = x_i$, $i = 1, \ldots, k$. We claim that $\{X_1(t_2), \ldots, X_k(t_2)\}$ is a basis for ker f_{*q_1}. First note that, since the parallel translation $P_{t_1}^{t_2}(\gamma) : (T_{p_1}M_1, g_{1p_1}) \to (T_{q_1}M_1, g_{1q_1})$ is an isometry, $\{X_1(t_2), \ldots, X_k(t_2)\}$ is a linearly independent set. Also, by Proposition 3.2.1, since f preserves parallel translation, f_*X_i $i = 1, \ldots, k$ are also parallel vector fields along the curve $f \circ \gamma$ with $(f_*X_i)(t_1) = 0$. By the uniqueness of parallel vector fields along a curve, (see Proposition 2.2.1), we have $(f_*X_i)(t_2) = 0$. Thus $\{X_1(t_2), \ldots, X_k(t_2)\}$ is a linearly independent set contained in ker f_{*q_1}, that is, dim (ker f_{*q_1}) $\geq k$. Also, by considering the same argument for the reverse curve to γ, it follows that dim (ker f_{*q_1}) $= k$.

(2) First note that, $\nabla f_* = 0$ if and only if $\nabla_x f_* = 0$ for every $x \in TM_1$. Hence by Lemma 3.1.1, since

$$(\overset{1}{\nabla} \|f_*\|^2)(p_1) = 2\sum_{i=1}^{n_1} g(x_i, x_i)(^*\left(\nabla_x f_*\right) \circ f_{*p_1}\big) x_i,$$

where $\{x_1, \ldots, x_{n_1}\}$ is an orthonormal basis for $(T_{p_1}M_1, g_{1p_1})$, we have $\overset{1}{\nabla} \|f_*\|^2 = 0$ on M_1. Thus, $\|f_*\|^2$ is constant on M_1. $\square$

Remark 3.2.1 Let $f : (M_1, g_1) \to (M_2, g_2)$ be an affine map, where M_1 is connected. Then since the rank of f is constant on M_1, it follows from the rank theorem for functions (cf. [AMR, p. 205], [BJ, p. 45], [F1, Theorem 4.3]) that $\ker f_*$ is an integrable subbundle of TM_1. Also since $f_*(\overset{1}{\nabla}_X Y) = \overset{2}{\nabla}_X f_*Y = 0$ for every X, $Y \in \Gamma \ker f_*$, $\overset{1}{\nabla}_X Y \in \Gamma \ker f_*$ for every X, $Y \in \Gamma \ker f_*$, that is, integral manifolds of $\ker f_*$ are totally geodesic in (M_1, g_1). Hence each integral manifold of $\ker f_*$ is either a semi-Riemannian or a degenerate submanifold of (M_1, g_1). (See Proposition A.1.2).

Now we define harmonic maps. However, we will not study harmonic maps as much as affine maps since we will only need them for some remarks on semi-Riemannian maps.

Definition 3.2.2 Let $f : (M_1, g_1) \to (M_2, g_2)$ be a map. Then the *tension field* $\tau(f)$ *of* f is defined to be the trace of ∇f_* with respect to g_1, that is,

$$\tau(f) = \sum_{i=1}^{n_1} g_1(X_i, X_i)(\nabla f_*)(X_i, X_i),$$

where $\{X_1, \ldots, X_{n_1}\}$ is an orthonormal local basis frame for TM_1.

Note that the tension field of a map $f : (M_1, g_1) \to (M_2, g_2)$ is a vector field along f, that is, $\tau(f) \in \Gamma_f TM_2$.

Definition 3.2.3 A map $f : (M_1, g_1) \to (M_2, g_2)$ is called *harmonic* if it has vanishing tension field, that is, $\tau(f) = 0$.

Remark 3.2.2 It is an instructive exercise to show that, if $f : (M, g) \to (\mathbf{R}, dt \otimes dt)$ is a map, then for every X, $Y \in \Gamma TM$,

$$(\nabla f_*)(X, Y) = H_f(X, Y)\frac{d}{dt} \circ f$$

and hence,

$$\tau(f) = -(\Delta f)\frac{d}{dt} \circ f.$$

Thus, f is affine if and only if $H_f = 0$, and f is harmonic if and only if f is a harmonic function on (M, g). (See section 2.4).

Proposition 3.2.5 *Let $I \subseteq \mathbf{R}$ be an open interval and (M, g) be a semi-Riemannian manifold. Then every harmonic map $f : (I, dt \otimes dt) \to (M, g)$ is an affine map and $f : I \to M$ is a geodesic of (M, g).*

Proof. Note that, since

$$\tau(f) = (\nabla f_*)(\frac{d}{dt}, \frac{d}{dt}),$$

if f is harmonic then f is affine. Also, since

$$0 = \tau(f) = (\nabla f_*)(\frac{d}{dt}, \frac{d}{dt}) = \nabla_{\frac{d}{dt}} f_* \frac{d}{dt} = \nabla_{\frac{d}{dt}} \dot{f},$$

f is a geodesic of (M, g), where ∇ also denotes the pullback of the Levi-Civita connection ∇ of (M, g) along f. □

3.3 The Generalized Semi-Riemannian Divergence Theorem

In this section we generalize the semi-Riemannian Divergence Theorem (Theorem 2.5.2) to vector fields along a map $f : (M_1, g_1) \to (M_2, g_2)$. For this, we will first define the divergence of a vector field along a map.

Let $f : (M_1, g_1) \to (M_2, g_2)$ be a map and Z be a vector field along f, that is, $Z \in \Gamma_f TM_2$. Now define a map

$$^*f_* \overset{2}{\nabla} Z : \Gamma TM_1 \to \Gamma TM_1$$

by

$$(^*f_* \overset{2}{\nabla} Z)(X) = {}^*f_{*\pi_1(X)} \overset{2}{\nabla}_X Z,$$

where $\pi_1 : TM_1 \to M_1$ is the canonical projection, $^*f_{*\pi_1(X)}$ is the adjoint of $f_{*\pi_1(X)}$: $(T_{\pi_1(X)}M_1, g_{1\pi_1(X)}) \to (T_{f(\pi_1(X))}M_1, g_{1f(\pi_1(X))})$ and $\overset{2}{\nabla}$ also denotes the pullback of the Levi-Civita connection $\overset{2}{\nabla}$ of (M_2, g_2) along f. Note that $^*f_* \overset{2}{\nabla} Z : \Gamma TM_1 \to \Gamma TM_1$ is a linear map for which the trace is well-defined.

Definition 3.3.1 Let $f : (M_1, g_1) \to (M_2, g_2)$ be a map and let Z be a vector field along f, that is, $Z \in \Gamma_f TM_2$. Then the *divergence of* Z is defined by

$$\text{div } Z = \text{ trace } ^*f_* \overset{2}{\nabla} Z.$$

Remark 3.3.1 Let $f : (M_1, g_1) \to (M_2, g_2)$ be a map and $Z \in \Gamma_f TM_2$. Then note that,

$$\begin{aligned}
\text{div } Z &= \text{trace } {}^*f_* \overset{2}{\nabla} Z \\
&= \sum_{i=1}^{n_1} g_1(X_i, X_i) g_1(({}^*f_* \overset{2}{\nabla} Z)(X_i), X_i) \\
&= \sum_{i=1}^{n_1} g_1(X_i, X_i) g_1({}^*f_{*\pi_1(X_i)} \overset{2}{\nabla}_{X_i} Z, X_i) \\
&= \sum_{i=1}^{n_1} g_1(X_i, X_i) g_2(\overset{2}{\nabla}_{X_i} Z, f_* X_i),
\end{aligned}$$

where $\{X_1, \ldots, X_{n_1}\}$ is an orthonormal local frame for TM_1. Also note that if we set $(M_1, g_1) = (M_2, g_2) = (M, g)$ and $f = \text{id}$ then the divergence of a vector field along f reduces to the usual divergence of a vector field on M. (See Definition 2.4.1).

Now following the notation and terminology of Section 2.5 of Chapter 2, we state the generalized semi-Riemannian divergence theorem as follows:

Theorem 3.3.1 [GK4] **(The Generalized Semi-Riemannian Divergence Theorem)** *Let (M_1, g_1) be an oriented semi-Riemannian manifold with boundary ∂M_1 (possibly $\partial M_1 = \emptyset$) and semi-Riemannian volume form μ_1, and let (M_2, g_2) be a semi-Riemannian manifold. Let $f : (M_1, g_1) \to (M_2, g_2)$ be a map and Z be a vector field along f with compact support. If $\partial(M_1)_0$ has measure zero in ∂M_1 then*

$$\begin{aligned}
\int_{M_1} (\text{div } Z)\mu_1 &+ \int_{M_1} g_2(Z, \tau(f))\mu_1 \\
&= \int_{\partial(M_1)_+} g_2(Z, f_* N_1)\mu_{1\partial(M_1)'} - \int_{\partial(M_1)_-} g_2(Z, f_* N_1)\mu_{1\partial(M_1)'},
\end{aligned}$$

where N_1 is the unit outward normal vector field to ∂M_1 defined on $\partial(M_1)'$.

Proof. Let ${}^*f_* Z$ be a vector field on M_1 defined by

$$({}^*f_* Z)(p_1) = {}^*f_{*p_1}(Z(p_1))$$

at each $p_1 \in M_1$, where ${}^*f_{*p_1}$ is the adjoint of f_{*p_1}. Then if $\{X_1, \ldots, X_{n_1}\}$ is an adapted moving frame near p_1, we have

$$
\begin{aligned}
(\operatorname{div} {}^*f_*Z)(p_1) &= \left[\sum_{i=1}^{n_1} g_1(X_i,X_i)g_1(\overset{1}{\nabla}_{X_i}({}^*f_*Z),X_i)\right](p_1) \\
&= \left[\sum_{i=1}^{n_1} g_1(X_i,X_i)X_i g_1({}^*f_*Z,X_i)\right](p_1) \\
&= \left[\sum_{i=1}^{n_1} g_1(X_i,X_i)X_i g_2(Z,f_*X_i)\right](p_1) \\
&= \left[\sum_{i=1}^{n_1} g_1(X_i,X_i)g_2(\overset{2}{\nabla}_{X_i} Z,f_*X_i)\right](p_1) \\
&\quad + \left[\sum_{i=1}^{n_1} g_1(X_i,X_i)g_2(Z,\overset{2}{\nabla}_{X_i} f_*X_i)\right](p_1) \\
&= \left[\sum_{i=1}^{n_1} g_1(X_i,X_i)g_2(\overset{2}{\nabla}_{X_i} Z,f_*X_i)\right](p_1) \\
&\quad + \left[\sum_{i=1}^{n_1} g_1(X_i,X_i)g_2(Z,(\nabla f_*)(X_i,X_i))\right](p_1) \\
&= [\operatorname{div} Z + g_2(Z,\tau(f))](p_1),
\end{aligned}
$$

where we also denoted the pullback of the Levi-Civita connection $\overset{2}{\nabla}$ of (M_2, g_2) along f by $\overset{2}{\nabla}$. Hence

$$\operatorname{div} {}^*f_*Z = \operatorname{div} Z + g_2(Z,\tau(f)).$$

Now by applying the semi-Riemannian Divergence Theorem (Theorem 2.5.2) to ${}^*f_*Z$ on M_1, we obtain

$$
\begin{aligned}
\int_{M_1} (\operatorname{div} {}^*f_*Z)\mu_1 &= \int_{\partial(M_1)_+} g_1({}^*f_*Z,N_1)\mu_{1\partial(M_1)'} - \int_{\partial(M_1)_-} g_1({}^*f_*Z,N_1)\mu_{1\partial(M_1)'} \\
&= \int_{\partial(M_1)_+} g_2(Z,f_*N_1)\mu_{1\partial(M_1)'} - \int_{\partial(M_1)_-} g_2(Z,f_*N_1)\mu_{1\partial(M_1)'}.
\end{aligned}
$$

Thus, since $\operatorname{div} {}^*f_*Z = \operatorname{div} Z + g_2(Z,\tau(f))$,

$$
\begin{aligned}
&\int_{M_1} (\operatorname{div} Z)\mu_1 + \int_{M_1} g_2(Z,\tau(f))\mu_1 \\
&\quad = \int_{\partial(M_1)_+} g_2(Z,f_*N_1)\mu_{1\partial(M_1)'} - \int_{\partial(M_1)_-} g_2(Z,f_*N_1)\mu_{1\partial(M_1)'}.
\end{aligned}
$$

□

Finally in this section, we will make an application of the generalized semi-Riemannian divergence theorem to harmonic maps.

Proposition 3.3.1 *Let (M_1, g_1) be an oriented compact semi-Riemannian manifold with semi-Riemannian volume form μ_1 and let (M_2, g_2) be a Riemannian manifold. If $f : (M_1, g_1) \to (M_2, g_2)$ is a map with $\int_{M_1} (\text{div}\ \tau(f))\mu_1 \geq 0$, then f is harmonic.*

Proof. Since $\partial M_1 = \emptyset$, by Theorem 3.3.1,

$$\int_{M_1} (\text{div}\ \tau(f))\mu_1 + \int_{M_1} g_2(\tau(f), \tau(f))\mu_1 = 0.$$

Thus since $\int_{M_1} (\text{div}\ \tau(f))\mu_1 \geq 0$ and g_2 is a positive definite metric tensor, $g_2(\tau(f), \tau(f)) = 0$ and hence, $\tau(f) = 0$. □

Note here that Proposition 3.3.1 fails to hold if (M_1, g_1) is not compact. Indeed, the map $f : (\mathbf{R}, dt \otimes dt) \to (\mathbf{R}, dt \otimes dt)$ given by $f(t) = \frac{1}{2}t^2$ has div $\tau(f) = 0$ but f is not harmonic. Also see [AK] for a generalization of Proposition 3.3.1.

3.4 The Bochner Identity

In this section, we derive an important identity for the maps between semi-Riemannian manifolds which will be used in showing that a Riemannian map is affine under certain circumferences.

Let $f : (M_1, g_1) \to (M_2, g_2)$ be a map. For each fixed $X \in \Gamma TM_1$, define a map

$$^*(\nabla_X f_*) : \Gamma_f TM_2 \to \Gamma TM_1$$

by

$$(^*(\nabla_X f_*)(Y))(p_1) = {}^*(\nabla_{X(p_1)} f_*)Y(p_1),$$

where $^*(\nabla_{X(p_1)} f_*)$ is the adjoint of $\nabla_{X(p_1)} f_*$. (See Section 3.1). Clearly $^*(\nabla_X f_*)$ is a linear map and by employing it in the expression of $(\overset{1}{\nabla} \|f_*\|^2)(p_1)$ (see Lemma 3.1.1), we obtain a local expression for $\overset{1}{\nabla} \|f_*\|^2$ as

$$\overset{1}{\nabla} \|f_*\|^2 = 2\sum_{i=1}^{n_1} g(X_i, X_i)(^*(\nabla_{X_i} f_*) \circ f_*)X_i,$$

where $\{X_1, \ldots, X_{n_1}\}$ is an orthonormal local frame for TM_1.

Lemma 3.4.1 (The Bochner Identity) *Let $f : (M_1, g_1) \to (M_2, g_2)$ be a map. Then*

$$\begin{aligned}
-\frac{1}{2}(\overset{1}{\Delta} \|f_*\|^2)(p_1) &= \sum_{i,j=1}^{n_1} g_1(x_i, x_i) g_1(x_j, x_j) g_2((\nabla f_*)(x_i, x_j), (\nabla f_*)(x_i, x_j)) \\
&\quad - \sum_{i,j=1}^{n_1} g_1(x_i, x_i) g_1(x_j, x_j) g_2(\overset{2}{R}(f_* x_i, f_* x_j) f_* x_j, f_* x_i) \\
&\quad + \sum_{i,j=1}^{n_1} g_1(x_i, x_i) g_1(x_j, x_j) g_1((^* f_{*p_1} \circ f_{*p_1}) x_i, \overset{1}{R}(x_i, x_j) x_j) \\
&\quad + (\text{div } \tau(f))(p_1),
\end{aligned}$$

where $\overset{1}{\Delta} \|f_*\|^2$ is the Laplacian of $\|f_*\|^2$ on (M_1, g_1), $\{x_1, \ldots, x_{n_1}\}$ is an orthonormal basis for $(T_{p_1} M_1, g_{1p_1})$ and, $\overset{1}{R}$ and $\overset{2}{R}$ are the curvature tensors of (M_1, g_1) and (M_2, g_2), respectively.

Proof. Let $\{x_1, \ldots, x_{n_1}\}$ be an orthonormal basis for $(T_{p_1} M_1, g_{1p_1})$ and let $\{X_1, \ldots, X_{n_1}\}$ be the extension of $\{x_1, \ldots, x_{n_1}\}$ to an adapted moving frame near p_1. Also let $\overset{1}{\nabla}$ be the Levi-Civita connection of (M_1, g_1) and let $\overset{2}{\nabla}$ also denote both the Levi-Civita connection of (M_2, g_2) and its pullback along f. By $\overset{2}{R}$ we will both denote the curvature tensors of the Levi-Civita connection of $\overset{2}{\nabla}$ of (M_2, g_2) and its pullback along f. Then by using Theorem 2.2.3, Proposition 2.3.1, Proposition 2.3.2 and Proposition 3.1.1,

$$\begin{aligned}
-\frac{1}{2}(\overset{1}{\Delta} \|f_*\|^2)(p_1) &= \frac{1}{2}(\text{div } \overset{1}{\nabla} \|f_*\|^2)(p_1) \\
&= \left[\text{div} \sum_{i=1}^{n_1} g_1(X_i, X_i)(^*(\nabla_{X_i} f_*) \circ f_*) X_i \right](p_1) \\
&= \left[\sum_{i,j=1}^{n_1} g_1(X_i, X_i) g_1(X_j, X_j) g_1(\overset{1}{\nabla}_{X_j} (^*(\nabla_{X_i} f_*) \circ f_*) X_i, X_j) \right](p_1) \\
&= \left[\sum_{i,j=1}^{n_1} g_1(X_i, X_i) g_1(X_j, X_j) X_j g_1((^*(\nabla_{X_i} f_*) \circ f_*) X_i, X_j) \right](p_1) \\
&= \left[\sum_{i,j=1}^{n_1} g_1(X_i, X_i) g_1(X_j, X_j) X_j g_2(f_* X_i, (\nabla_{X_i} f_*) X_j) \right](p_1)
\end{aligned}$$

Thus,

$$
\begin{aligned}
-\frac{1}{2}(\overset{1}{\Delta}\|f_*\|^2)(p_1) &= \left[\sum_{i,j=1}^{n_1} g_1(X_i,X_i)g_1(X_j,X_j)g_2(\overset{2}{\nabla}_{X_j}\, f_*X_i,(\nabla_{X_i}f_*)X_j)\right](p_1)\\
&+\left[\sum_{i,j=1}^{n_1} g_1(X_i,X_i)g_1(X_j,X_j)g_2(f_*X_i,\overset{2}{\nabla}_{X_j}\,((\nabla_{X_i}f_*)X_j))\right](p_1)\\
&=\left[\sum_{i,j=1}^{n_1} g_1(X_i,X_i)g_1(X_j,X_j)g_2((\nabla_{X_j}f_*)X_i,(\nabla_{X_i}f_*)X_j)\right](p_1)\\
&+\left[\sum_{i,j=1}^{n_1} g_1(X_i,X_i)g_1(X_j,X_j)g_2(f_*X_i,\overset{2}{\nabla}_{X_j}\overset{2}{\nabla}_{X_i}\, f_*X_j)\right](p_1)\\
&-\left[\sum_{i,j=1}^{n_1} g_1(X_i,X_i)g_1(X_j,X_j)g_2(f_*X_i,\overset{2}{\nabla}_{X_j}\,(f_*\,\overset{1}{\nabla}_{X_i}\,X_j)\right](p_1)\\
&=\left[\sum_{i,j=1}^{n_1} g_1(X_i,X_i)g_1(X_j,X_j)g_2((\nabla f_*)(X_j,X_i),(\nabla f_*)(X_i,X_j))\right](p_1)\\
&+\left[\sum_{i,j=1}^{n_1} g_1(X_i,X_i)g_1(X_j,X_j)g_2(f_*X_i,\overset{2}{R}\,(X_j,X_i)f_*X_j)\right](p_1)\\
&+\left[\sum_{i,j=1}^{n_1} g_1(X_i,X_i)g_1(X_j,X_j)g_2(f_*X_i,\overset{2}{\nabla}_{X_i}\overset{2}{\nabla}_{X_j}\, f_*X_j)\right](p_1)\\
&-\left[\sum_{i,j=1}^{n_1} g_1(X_i,X_i)g_1(X_j,X_j)g_2(f_*X_i,f_*\,\overset{1}{\nabla}_{X_j}\overset{1}{\nabla}_{X_i}\,X_j)\right](p_1)\\
&=\left[\sum_{i,j=1}^{n_1} g_1(X_i,X_i)g_1(X_j,X_j)g_2((\nabla f_*)(X_i,X_j),(\nabla f_*)(X_i,X_j))\right](p_1)\\
&+\left[\sum_{i,j=1}^{n_1} g_1(X_i,X_i)g_1(X_j,X_j)g_2(f_*X_i,\overset{2}{R}\,(f_*X_j,f_*X_i)f_*X_j)\right](p_1)\\
&+\left[\sum_{i,j=1}^{n_1} g_1(X_i,X_i)g_1(X_j,X_j)X_ig_2(f_*X_i,\overset{2}{\nabla}_{X_j}\, f_*X_j)\right](p_1)
\end{aligned}
$$

$$-\left[\sum_{i,j=1}^{n_1} g_1(X_i,X_i)g_1(X_j,X_j)g_2(\overset{2}{\nabla}_{X_i} f_*X_i, \overset{2}{\nabla}_{X_j} f_*X_j)\right](p_1)$$

$$-\left[\sum_{i,j=1}^{n_1} g_1(X_i,X_i)g_1(X_j,X_j)g_2(f_*X_i, f_* \overset{1}{\nabla}_{X_j}\overset{1}{\nabla}_{X_i} X_j)\right](p_1).$$

Then, one has

$$-\frac{1}{2}(\overset{1}{\Delta} \|f_*\|^2)(p_1)$$

$$= \left[\sum_{i,j=1}^{n_1} g_1(X_i,X_i)g_1(X_j,X_j)g_2((\nabla f_*)(X_i,X_j),(\nabla f_*)(X_i,X_j))\right](p_1)$$

$$-\left[\sum_{i,j=1}^{n_1} g_1(X_i,X_i)g_1(X_j,X_j)g_2(\overset{2}{R}(f_*X_i,f_*X_j)f_*X_j, f_*X_i)\right](p_1)$$

$$+\left[\sum_{i,j=1}^{n_1} g_1(X_i,X_i)g_1(X_j,X_j)X_i g_2(f_*X_i,(\nabla_{X_j}f_*)X_j)\right](p_1)$$

$$+\left[\sum_{i,j=1}^{n_1} g_1(X_i,X_i)g_1(X_j,X_j)X_i g_2(f_*X_i, f_* \overset{1}{\nabla}_{X_j} X_j)\right](p_1)$$

$$-\left[\sum_{i,j=1}^{n_1} g_1(X_i,X_i)g_1(X_j,X_j)g_2((\nabla_{X_i}f_*)X_i,(\nabla_{X_j}f_*)X_j)\right](p_1)$$

$$-\left[\sum_{i,j=1}^{n_1} g_1(X_i,X_i)g_1(X_j,X_j)g_2(f_*X_i, f_* \overset{1}{R}(X_j,X_i)X_j)\right](p_1)$$

$$-\left[\sum_{i,j=1}^{n_1} g_1(X_i,X_i)g_1(X_j,X_j)g_2(f_*X_i, f_* \overset{1}{\nabla}_{X_i}\overset{1}{\nabla}_{X_j} X_j)\right](p_1).$$

Now, it follows that

$$
\begin{aligned}
-\frac{1}{2}(\overset{1}{\Delta}\, \|f_*\|^2)(p_1)
&= \left[\sum_{i,j=1}^{n_1} g_1(X_i,X_i)g_1(X_j,X_j)g_2((\nabla f_*)(X_i,X_j),(\nabla f_*)(X_i,X_j))\right](p_1)\\
&\quad - \left[\sum_{i,j=1}^{n_1} g_1(X_i,X_i)g_1(X_j,X_j)g_2(\overset{2}{R}\,(f_*X_i,f_*X_j)f_*X_j,f_*X_i)\right](p_1)\\
&\quad + \left[\sum_{i,j=1}^{n_1} g_1(X_i,X_i)g_1(X_j,X_j)g_2(\overset{2}{\nabla}_{X_i}\, f_*X_i,(\nabla_{X_j}f_*)X_j)\right](p_1)\\
&\quad + \left[\sum_{i,j=1}^{n_1} g_1(X_i,X_i)g_1(X_j,X_j)g_2(f_*X_i,\overset{2}{\nabla}_{X_i}\,((\nabla_{X_j}f_*)X_j))\right](p_1)\\
&\quad + \left[\sum_{i,j=1}^{n_1} g_1(X_i,X_i)g_1(X_j,X_j)g_2(f_*X_i,\overset{2}{\nabla}_{X_i}\,(f_*\,\overset{1}{\nabla}_{X_j}\,X_j))\right](p_1)\\
&\quad - \left[\sum_{i,j=1}^{n_1} g_1(X_i,X_i)g_1(X_j,X_j)g_2((\nabla_{X_i}f_*)X_i,(\nabla_{X_j}f_*)X_j)\right](p_1)\\
&\quad + \left[\sum_{i,j=1}^{n_1} g_1(X_i,X_i)g_1(X_j,X_j)g_2(f_*X_i,f_*\,\overset{1}{R}\,(X_i,X_j)X_j)\right](p_1)\\
&\quad - \left[\sum_{i,j=1}^{n_1} g_1(X_i,X_i)g_1(X_j,X_j)g_2(f_*X_i,f_*\,\overset{1}{\nabla}_{X_i}\overset{1}{\nabla}_{X_j}\,X_j)\right](p_1)\\
&= \left[\sum_{i,j=1}^{n_1} g_1(X_i,X_i)g_1(X_j,X_j)g_2((\nabla f_*)(X_i,X_j),(\nabla f_*)(X_i,X_j))\right](p_1)\\
&\quad - \left[\sum_{i,j=1}^{n_1} g_1(X_i,X_i)g_1(X_j,X_j)g_2(\overset{2}{R}\,(f_*X_i,f_*X_j)f_*X_j,f_*X_i)\right](p_1)\\
&\quad + \left[\sum_{i,j=1}^{n_1} g_1(X_i,X_i)g_1(X_j,X_j)g_2((\nabla_{X_i}f_*)X_i,(\nabla_{X_j}f_*)X_j)\right](p_1)\\
&\quad + \left[\sum_{i,j=1}^{n_1} g_1(X_i,X_i)g_1(X_j,X_j)g_2(f_*X_i,\overset{2}{\nabla}_{X_i}\,((\nabla_{X_j}f_*)X_j))\right](p_1)
\end{aligned}
$$

$$+\left[\sum_{i,j=1}^{n_1} g_1(X_i,X_i)g_1(X_j,X_j)g_2(f_*X_i, f_* \overset{1}{\nabla}_{X_i}\overset{1}{\nabla}_{X_j} X_j)\right](p_1)$$

$$-\left[\sum_{i,j=1}^{n_1} g_1(X_i,X_i)g_1(X_j,X_j)g_2((\nabla_{X_i}f_*)X_i, (\nabla_{X_j}f_*)X_j)\right](p_1)$$

$$+\left[\sum_{i,j=1}^{n_1} g_1(X_i,X_i)g_1(X_j,X_j)g_2(f_*X_i, f_* \overset{1}{R}(X_i,X_j)X_j)\right](p_1)$$

$$-\left[\sum_{i,j=1}^{n_1} g_1(X_i,X_i)g_1(X_j,X_j)g_2(f_*X_i, f_* \overset{1}{\nabla}_{X_i}\overset{1}{\nabla}_{X_j} X_j)\right](p_1)$$

$$=\left[\sum_{i,j=1}^{n_1} g_1(X_i,X_i)g_1(X_j,X_j)g_2((\nabla f_*)(X_i,X_j), (\nabla f_*)(X_i,X_j))\right](p_1)$$

$$-\left[\sum_{i,j=1}^{n_1} g_1(X_i,X_i)g_1(X_j,X_j)g_2(\overset{2}{R}(f_*X_i, f_*X_j)f_*X_j, f_*X_i)\right](p_1)$$

$$+\left[\sum_{i,j=1}^{n_1} g_1(X_i,X_i)g_1(X_j,X_j)g_2(f_*X_i, \overset{2}{\nabla}_{X_i}((\nabla f_*)(X_j,X_j)))\right](p_1)$$

$$+\left[\sum_{i,j=1}^{n_1} g_1(X_i,X_i)g_1(X_j,X_j)g_2(f_*X_i, f_* \overset{1}{R}(X_i,X_j)X_j)\right](p_1)$$

$$=\sum_{i,j=1}^{n_1} g_1(x_i,x_i)g_1(x_j,x_j)g_2((\nabla f_*)(x_i,x_j), (\nabla f_*)(x_i,x_j))$$

$$-\sum_{i,j=1}^{n_1} g_1(x_i,x_i)g_1(x_j,x_j)g_2(\overset{2}{R}(f_*x_i, f_*x_j)f_*x_j, f_*x_i)$$

$$+(\operatorname{div} \tau(f))(p_1)$$

$$+\sum_{i,j=1}^{n_1} g_1(x_i,x_i)g_1(x_j,x_j)g_2(({}^*f_{*p_1} \circ f_{*p_1})x_i, \overset{1}{R}(x_i,x_j)x_j).$$

□

We also have the following corollary to the Lemma above.

Corollary 3.4.1 *Let $f : (M, g) \to \mathbf{R}$ be a map. Then*

$$-\frac{1}{2}\Delta g(\nabla f, \nabla f) = \|h_f\|^2 + Ric(\nabla f, \nabla f) - g(\nabla \Delta f, \nabla f),$$

where ∇f and $\Delta g(\nabla f, \nabla f)$ denote the gradient and Laplacian of f and $g(\nabla f, \nabla f)$ on (M, g), respectively.

Proof. If we set $(M_1, g_1) = (M, g)$ and $(M_2, g_2) = (\mathbf{R}, dt \otimes dt)$ in Lemma 3.4.1, by observing that, for $X \in \Gamma TM$,

$$f_* X = g(\nabla f, X)\frac{d}{dt} \circ f,$$

it can be shown that,

$$\|f_*\|^2(p) = g(\nabla f, \nabla f)(p),$$

$$\sum_{i,j=1}^{n_1} g_1(x_i, x_i) g_1(x_j, x_j) g_2((\nabla f_*)(x_i, x_j), (\nabla f_*)(x_i, x_j)) = \|h_f\|^2(p),$$

$$\sum_{i,j=1}^{n_1} g_1(x_i, x_i) g_1(x_j, x_j) g_2(\overset{2}{R}(f_* x_i, f_* x_j) f_* x_j, f_* x_i) = 0,$$

$$\sum_{i,j=1}^{n_1} g_1(x_i, x_i) g_1(x_j, x_j) g_2(({}^* f_{*p_1} \circ f_{*p_1}) x_i, \overset{1}{R}(x_i, x_j) x_j) = Ric(\nabla f, \nabla f)(p),$$

$$(\text{div } \tau(f))(p_1) = -g(\nabla \Delta f, \nabla f)(p),$$

where $p_1 = p \in M_1 = M$. Hence by Lemma 3.4.1, the claim follows. We leave these computations as an instructive exercise to the reader and compute $-\frac{1}{2}\Delta g(\nabla f, \nabla f)$ directly which is much shorter. Let $\{X_1, \ldots, X_n\}$ be an adapted moving frame near $p \in M$. Then by Corollary 3.1.1,

$$\begin{aligned} -\frac{1}{2}(\Delta g(\nabla f, \nabla f))(p) &= \frac{1}{2}(\text{div } \nabla g(\nabla f, \nabla f))(p) \\ &= (\text{div } \nabla_{\nabla f} \nabla f)(p) \\ &= \left[\sum_{i=1}^{n} g(X_i, X_i) g(\nabla_{X_i} \nabla_{\nabla f} \nabla f, X_i)\right](p). \end{aligned}$$

Thus,

$$\begin{aligned}
-\frac{1}{2}(\Delta g(\nabla f,\nabla f))(p) &= \left[\sum_{i=1}^{n} g(X_i,X_i)g(R(X_i,\nabla f)\nabla f,X_i)\right](p)\\
&+\left[\sum_{i=1}^{n} g(X_i,X_i)g(\nabla_{\nabla f}\nabla_{X_i}\nabla f,X_i)\right](p)\\
&+\left[\sum_{i=1}^{n} g(X_i,X_i)g(\nabla_{[X_i,\nabla f]}\nabla f,X_i)\right](p)\\
&= \left[\sum_{i=1}^{n} g(X_i,X_i)g(R(X_i,\nabla f)\nabla f,X_i)\right](p)\\
&+\left[\sum_{i=1}^{n} g(X_i,X_i)(\nabla f)g(\nabla_{X_i}\nabla f,X_i)\right](p)\\
&+\left[\sum_{i=1}^{n} g(X_i,X_i)g(\nabla_{\nabla_{X_i}\nabla f}\nabla f,X_i)\right](p)\\
&= Ric(\nabla f(p),\nabla f(p)) - ((\nabla f)\Delta f)(p)\\
&+\left[\sum_{i=1}^{n} g(X_i,X_i)g(\nabla_{X_i}\nabla f,\nabla_{X_i}\nabla f)\right](p)\\
&= \left[Ric(\nabla f,\nabla f) - g(\nabla\Delta f,\nabla f) + \|h_f\|^2\right](p).
\end{aligned}$$

Hence

$$-\frac{1}{2}\Delta g(\nabla f,\nabla f) = \|h_f\|^2 + Ric(\nabla f,\nabla f) - g(\nabla\Delta f,\nabla f).$$

□

Finally in this section, we make two applications of the Bochner identity to show that some maps satisfying certain conditions are affine. Similar applications will also be made for Riemannian maps.

As a terminology, we call a semi-Riemannian manifold (M,g) of *nonnegative Ricci curvature* if the Ricci curvature of (M,g) with respect to every nonzero, nonnull vector $v \in TM$ is nonnegative. (See Definition 2.3.5). Also a semi-Riemannian manifold (M,g) is called of *nonpositive sectional curvature* if the sectional curvature of every nondegenerate plane in T_pM is nonpositive at each $p \in M$.

Also the following observation will be used in the proof of the theorem below. Let $f : (M_1,g_1) \to (M_2,g_2)$ be a map. Then note that ${}^*f_{*p_1} \circ f_{*p_1} : (T_{p_1}M_1, g_{1p_1}) \to$

$(T_{p_1}M_1, g_{1p_1})$ is a self-adjoint linear transformation. (See section 1.2). Moreover, if (M_2, g_2) is a Riemannian manifold, then ${}^*f_{*p_1} \circ f_{*p_1}$ is a nonnegative linear transformation on $(T_{p_1}M_1, g_{1p_1})$, that is, $g_1(({}^*f_{*p_1} \circ f_{*p_1})x, x) \geq 0$ for every $x \in T_pM$. Hence, in particular, if (M_1, g_1) and (M_2, g_2) are Riemannian manifolds then ${}^*f_{*p_1} \circ f_{*p_1}$ is diagonalizable with nonnegative eigenvalues.

Theorem 3.4.1 *Let $f : (M_1, g_1) \to (M_2, g_2)$ be a map, where (M_1, g_1) is an oriented compact Riemannian manifold with Riemannian volume form μ_1 and (M_2, g_2) is a Riemannian manifold. Assume that (M_1, g_1) is of nonnegative Ricci curvature and (M_2, g_2) is of nonpositive sectional curvature. If $\int_{M_1} (\text{div } \tau(f))\mu_1 \geq 0$, then f is affine.*

Proof. First observe that, since $\int_{M_1} (\text{div } \tau(f))\mu_1 \geq 0$, by Proposition 3.3.1, f is harmonic and hence div $\tau(f) = 0$. Now let $\{x_1, \ldots, x_{n_1}\}$ be an orthonormal basis for $(T_{p_1}M_1, g_{1p_1})$ consisting of unit eigenvectors of ${}^*f_{*p_1} \circ f_{*p_1}$ with the corresponding set of (nonnegative) eigenvalues $\{\lambda_1, \ldots, \lambda_{n_1}\}$. Then with respect to this basis, we have by the Bochner identity (Lemma 3.4.1) that

$$\begin{aligned} -\frac{1}{2}(\overset{1}{\Delta} \|f_*\|^2)(p_1) &= \sum_{i,j=1}^{n_1} g_2((\nabla f_*)(x_i, x_j), (\nabla f_*)(x_i, x_j)) \\ &\quad - \sum_{i,j=1}^{n_1} g_2(\overset{2}{R}(f_*x_i, f_*x_j)f_*x_j, f_*x_i) \\ &\quad + \sum_{i,j=1}^{n_1} \lambda_i g_1(x_i, \overset{1}{R}(x_i, x_j)x_j) \end{aligned}$$

and thus,

$$\begin{aligned} -\frac{1}{2}(\overset{1}{\Delta} \|f_*\|^2)(p_1) &= \sum_{i,j=1}^{n_1} g_2((\nabla f_*)(x_i, x_j), (\nabla f_*)(x_i, x_j)) \\ &\quad - \sum_{i,j=1}^{n_1} g_2(\overset{2}{R}(f_*x_i, f_*x_j)f_*x_j, f_*x_i) \\ &\quad + \sum_{i=1}^{n_1} \lambda_i r_1(x_i), \end{aligned}$$

where $r_1(x_i)$ is the Ricci curvature of (M_1, g_1) with respect to $x_i \in T_{p_1}M_1$. Thus, since g_2 is a positive definite metric tensor, (M_1, g_1) is of nonnegative Ricci curvature and (M_2, g_2) is of nonpositive sectional curvature,

$$-\frac{1}{2}(\overset{1}{\Delta}\,\|f_*\|^2)(p_1) \geq 0$$

at each $p_1 \in M_1$. That is, $\|f_*\|^2$ is a subharmonic map on (M_1, g_1) and it follows from Theorem 2.5.4 that $\|f_*\|^2$ is harmonic on (M_1, g_1), that is, $\overset{1}{\Delta}\,\|f_*\|^2 = 0$. Thus we have

$$\begin{aligned} 0 &= \sum_{i,j=1}^{n_1} g_2((\nabla f_*)(x_i, x_j), (\nabla f_*)(x_i, x_j)) \\ &\quad - \sum_{i,j=1}^{n_1} g_2(\overset{2}{R}\,(f_*x_i, f_*x_j)f_*x_j, f_*x_i) \\ &\quad + \sum_{i=1}^{n_1} \lambda_i r_1(x_i). \end{aligned}$$

Hence, again since (M_1, g_1) is of nonnegative Ricci curvature and (M_2, g_2) is of nonpositive sectional curvature, we necessarily have

$$g_2((\nabla f_*)(x_i, x_j), (\nabla f_*)(x_i, x_j)) = 0$$

for each $1 \leq i, j \leq n_1$. Thus it follows that $(\nabla f_*)_{p_1} = 0$ at each $p_1 \in M_1$ since g_2 is a positive definite metric tensor. That is, f is an affine map. □

In the theorem above, the curvature restriction on (M_2, g_2) may be removed by putting a restriction to the rank of f.

Theorem 3.4.2 *Let $f : (M_1, g_1) \to (M_2, g_2)$ be a map, where (M_1, g_1) is an oriented compact Riemannian manifold with Riemannian volume form μ_1 and (M_2, g_2) is a Riemannian manifold. Assume that (M_1, g_1) is of nonnegative Ricci curvature and* $\operatorname{rank}_{p_1}(f) \leq 1$ *at each $p_1 \in M_1$. If $\int_{M_1} (\operatorname{div} \tau(f))\mu_1 \geq 0$ then, f is affine.*

Proof. It can be shown as in the proof of Theorem 3.4.1 that f is harmonic and hence $\operatorname{div} \tau(f) = 0$. Also since $\operatorname{rank}_{p_1} f \leq 1$ at each $p_1 \in M_1$,

$$g_2(\overset{2}{R}\,(f_*x_i, f_*x_j)f_*x_j, f_*x_i) = 0$$

in the proof of Theorem 3.4.1. Hence it follows from the Bochner identity (Lemma 3.4.1) as in the proof of Theorem 3.4.1 that $\|f_*\|^2$ is harmonic on (M_1, g_1), that is, $\overset{1}{\Delta}\,\|f_*\|^2 = 0$. Thus the Bochner identity in the proof of Theorem 3.4.1 becomes

$$
\begin{aligned}
0 &= \sum_{i,j=1}^{n_1} g_2((\nabla f_*)(x_i, x_j), (\nabla f_*)(x_i, x_j)) \\
&\quad + \sum_{i=1}^{n_1} \lambda_i r_1(x_i)
\end{aligned}
$$

where $\{x_1, \ldots, x_{n_1}\}$ is an orthonormal basis for $(T_{p_1}M_1, g_{1p_1})$ consisting of unit eigenvectors of ${}^*f_{*p_1} \circ f_{*p_1}$ with the corresponding set of (nonnegative) eigenvalues $\{\lambda_1, \ldots, \lambda_{n_1}\}$. Hence since (M_1, g_1) is of nonnegative Ricci curvature, it follows as in the proof of Theorem 3.4.1 that f is affine. □

Note here that Theorems 3.4.1 and 3.4.2 are generalizations of two well-known results about harmonic maps where f is assumed to be harmonic. (See [EL, p. 13]). Hence the above theorems are stronger because of their weaker assumption $\int_{M_1} (\operatorname{div} \tau(f))\mu_1 \geq 0$ rather than $\tau(f) = 0$.

Chapter 4

Semi-Riemannian Maps

In this chapter we introduce the notion of semi-Riemannian maps. Intuitively a semi-Riemannian map is a map between semi-Riemannian manifolds which is as "isometric as it can be." Hence, as one expects, the existence of a semi-Riemannian map between semi-Riemannian manifolds enables us to compare some geometric properties of the semi-Riemannian manifolds in its domain and range. Clearly, this concept is a generalization of the notion of isometry.

In section 4.1, we define the nondegenerate rank of a map between semi-Riemannian manifolds. Also we introduce the generalized eikonal equation which states a relation between the square norm of the tangent map and the nondegenerate rank of a map between semi-Riemannian manifolds. Then we define semi-Riemannian maps and observe that semi-Riemannian maps are solutions of the generalized eikonal equation. In section 4.2, we make our first application of semi-Riemannian maps to a generalization of semi-Riemannian submersions so as to admit degenerate submanifolds as fibers.

As a special case, a semi-Riemannian map between Riemannian manifolds is called a Riemannian map. In comparison to semi-Riemannian maps between semi-Riemannian manifolds, a Riemannian map behaves more properly. We devote section 4.3 to a discussion of these special properties of Riemannian maps. Finally in section 4.4, we investigate the curvature conditions under which a Riemannian map becomes affine. As we discuss there, affine Riemannian maps may be used for comparison of sectional curvatures of the manifolds.

Throughout this chapter, let (M_1, g_1) and (M_2, g_2) denote semi-Riemannian manifolds of indices ν_1 and ν_2, and dimensions n_1 and n_2, respectively, unless otherwise stated. Also let $\overset{1}{\nabla}$ and $\overset{2}{\nabla}$ always denote the Levi-Civita connections of (M_1, g_1) and (M_2, g_2), respectively.

4.1 The Generalized Eikonal Equation and Semi-Riemannian Maps

Let $f : (M_1, g_1) \to (M_2, g_2)$ be a map and let $p_2 = f(p_1)$ for each $p_1 \in M_1$. Note that

$$f_{*p_1} : (T_{p_1}M_1, g_{1p_1}) \to (T_{p_2}M_2, g_{2p_2})$$

is a linear transformation between the inner product spaces $(T_{p_1}M_1,\ g_{1p_1})$ and $(T_{p_2}M_2,\ g_{2p_2})$. As in Section 1.2, set

$$\begin{aligned}
\mathcal{V}(p_1) &= \ker f_{*p_1} \\
\mathcal{H}(p_1) &= (\ker f_{*p_1})^{\perp} = \mathcal{V}(p_1)^{\perp} \\
L_1(p_1) &= (\ker f_{*p_1}) \cap (\ker f_{*p_1})^{\perp} = L_{g_{1p_1 \mathcal{V}(p_1)}} = L_{g_{1p_1 \mathcal{H}(p_1)}} \\
A_1(p_1) &= (\ker f_{*p_1}) + (\ker f_{*p_1})^{\perp} = L_1(p_1)^{\perp} \\
A_2(p_2) &= \text{range } f_{*p_1} \\
L_2(p_2) &= (\text{range } f_{*p_1}) \cap (\text{range } f_{*p_1})^{\perp} = L_{g_{2p_2 A_2(p_2)}}.
\end{aligned}$$

Also let $(\overline{\mathcal{H}}(p_1), \overline{g_1}_{p_1 \overline{\mathcal{H}}(p_1)})$ and $(\overline{A_2}(p_2), \overline{g_2}_{p_2 \overline{A_2}(p_2)})$ be the quotient inner product spaces introduced in section 1.2 and let

$$\overline{f_{*p_1}} : (\overline{\mathcal{H}}(p_1), \overline{g_1}_{p_1 \overline{\mathcal{H}}(p_1)}) \to (\overline{A_2}(p_2), \overline{g_2}_{p_2 \overline{A_2}(p_2)})$$

be the quotient of f_{*p_1}. (Cf. section 1.2).

Here it is worth emphasizing that $\overline{f_{*p_1}}$ may be neither injective or surjective. In fact, rank $\overline{f_{*p_1}} \leq$ rank f_{*p_1} $-$ dim $L_1(p_1)$. (See Remark 1.2.1).

Definition 4.1.1 Let $f : (M_1, g_1) \to (M_2, g_2)$ be a map. The *nondegenerate rank* of f at $p_1 \in M_1$ is defined to be the rank of $\overline{f_{*p_1}}$.

Definition 4.1.2 Let $f : (M_1, g_1) \to (M_2, g_2)$ be a map. Then f is called *weakly semi-Riemannian at* $p_1 \in M_1$ if $\|f_{*p_1}\|^2 =$ rank $\overline{f_{*p_1}}$, and f is called *weakly semi-Riemannian* if f is weakly semi-Riemannian at each $p_1 \in M_1$.

Remark 4.1.1 Note that, f is weakly semi-Riemannian at $p_1 \in M_1$ is equivalent to f_{*p_1} is weakly semi-Euclidean. (See Definition 1.2.5).

Remark 4.1.2 Let $f : (M_1, g_1) \to (M_2, g_2)$ be a weakly semi-Riemannian map. Then note that since the map $\|f_*\|^2 : M_1 \to \mathbf{R}$ defined by $\|f_*\|^2(p_1) = \|f_{*p_1}\|^2$ is a continuous function, the nondegenerate rank of f is constant on each connected component of M_1, that is, rank $\overline{f_{*p_1}}$ does not change on each connected component of M_1.

Definition 4.1.3 Let (M_1, g_1) and (M_2, g_2) be semi-Riemannian manifolds. Then the equation

$$\|f_*\|^2 = \operatorname{rank} \overline{f_*}$$

defined for the maps $f : (M_1, g_1) \to (M_2, g_2)$ by

$$\|f_*\|^2(p_1) = \operatorname{rank} \overline{f_{*p_1}}$$

at each $p_1 \in M_1$ is called the *generalized eikonal equation.* Hence weakly semi-Riemannian maps are the solutions of the generalized eikonal equation.

Now we define semi-Riemannian maps which form a certain subset of the solutions of the generalized eikonal equation with geometric nature.

Definition 4.1.4 Let $f : (M_1, g_1) \to (M_2, g_2)$ be a map. Then f is called *semi-Riemannian at* $p_1 \in M_1$ if

$$\overline{f_{*p_1}} : (\overline{\mathcal{H}}(p_1), \overline{g_1}_{p_1 \overline{\mathcal{H}}(p_1)}) \to (\overline{A_2}(p_2), \overline{g_2}_{p_2 \overline{A_2}(p_2)})$$

is an (into) isometry. Also f is called *semi-Riemannian* if f is semi-Riemannian at each $p_1 \in M_1$.

Remark 4.1.3 Note that, f is semi-Riemannian at $p_1 \in M_1$ is equivalent to f_{*p_1} is semi-Euclidean. (See Definition 1.2.4).

Theorem 4.1.1 *If $f : (M_1, g_1) \to (M_2, g_2)$ is semi-Riemannian at $p_1 \in M_1$ then f is weakly semi-Riemannian at $p_1 \in M_1$. Hence, if f is semi-Riemannian then f is weakly semi-Riemannian.*

Proof. Since f_{*p_1} is semi-Euclidean by Remark 4.1.3, it follows from Theorem 1.2.2 that $\|f_{*p_1}\|^2 = \operatorname{rank} \overline{f_{*p_1}}$, that is, f is weakly semi-Riemannian at p_1. □

The converse of the above theorem is not true in general as we know from section 1.2 that, a weakly semi-Euclidean linear transformation may not be semi-Euclidean.

Remark 4.1.4 Let $f : (M_1, g_1) \to (M_2, g_2)$ be a semi-Riemannian map at $p_1 \in M_1$. Then it is worth emphasizing that $\overline{f_{*p_1}}$ is injective (yet may not be surjective) and hence, rank $\overline{f_{*p_1}}$ = rank f_{*p_1} − dim $L_1(p_1)$. (See Remark 1.2.3). Also, if f is semi-Riemannian and M_1 is connected then, since f is weakly semi-Riemannian, it follows from Remark 4.1.2 that the nondegenerate rank of f is constant. But rank of f may change from point to point.

4.2 Semi-Riemannian Submersions

In the usual definition of semi-Riemannian submersion, the fibers of submersion are assumed to be semi-Riemannian submanifolds (cf. [O, p. 212]). Note that, semi-Riemannian submersions defined this way are semi-Riemannian maps. But the concept of semi-Riemannian maps is more general and can be used to generalize the usual definition of semi-Riemannian submersion which admit degenerate submanifolds as fibers.

Definition 4.2.1 A map $f : (M_1, g_1) \to (M_2, g_2)$ is called a *semi-Riemannian submersion* if f is a submersion and a semi-Riemannian map.

In the next remark, we disclose the content of this definition. But at this point, the reader is recommended to read Appendix A.2 on degenerate submanifolds. However for convenience, we will briefly explain what a degenerate submanifold is.

Let H be a submanifold of a semi-Riemannian manifold (M, g) and let g_H be the restriction of g to H. If (T_pH, g_{H_p}) is a degenerate subspace of (T_pM, g_p) of type (μ', ν', η') for every $p \in H$ then (H, g) is called a *degenerate submanifold of* (M, g) *of type* (μ', ν', η'). For a degenerate submanifold (H, g_H) of type (μ', ν', η'), the *degenerate subbundle* of (H, g_H) is defined by $L_{g_H} = \bigcup_{p \in H} L_{g_{H_p}}$, where $L_{g_{H_p}}$ is the degenerate space of g_{H_p} in T_pH. (See Definition 1.1.1). Then note that, by Remark 1.1.1, $L_{g_H} = TH \cap TH^\perp$, where TH is the tangent bundle of H and $TH^\perp$ is the orthogonal bundle to H defined as usual by $TH^\perp = \bigcup_{p \in M} T_pH^\perp$.

Remark 4.2.1 Let $f : (M_1, g_1) \to (M_2, g_2)$ be a semi-Riemannian submersion, where M_1 is connected. First note that, since f is a submersion,

$$\mathcal{V} = \bigcup_{p_1 \in M_1} \mathcal{V}(p_1) = \ker f_*$$

is a subbundle of TM_1. (In general, ker f_* is not necessarily a subbundle of TM_1 if f is only assumed to be a semi-Riemannian map). Hence

$$\mathcal{H} = \bigcup_{p_1 \in M_1} \mathcal{H}(p_1) = (\ker f_*)^\perp$$

is also a subbundle of TM_1, where $(\ker f_*)^\perp$ is the orthogonal bundle to $\ker f_*$ defined as usual by $(\ker f_*)^\perp = \bigcup_{p_1 \in M_1} (\ker f_{*p_1})^\perp$. Now observe that, since f is a semi-Riemannian submersion and M_1 is connected, dim $L_1(p_1) = \text{rank } f_{*p_1} - \text{rank } \overline{f_{*p_1}}$ is constant for all $p_1 \in M_1$. Thus

$$L_1 = \bigcup_{p_1 \in M_1} L_1(p_1) = \mathcal{V} \cap \mathcal{H} = (\ker f_*) \cap (\ker f_*)^\perp$$

is a subbundle of TM_1.

Also observe that, since L_1 is a subbundle of TM_1,

$$\overline{\mathcal{H}} = \bigcup_{p_1 \in M_1} \overline{\mathcal{H}}(p_1)$$

is a vector bundle over M_1 and, by Theorem 2.1.2, the inner products $\overline{g_1}_{p_1 \overline{\mathcal{H}}(p_1)}$ defined on $\overline{\mathcal{H}}(p_1)$ at each $p_1 \in M_1$ define a metric tensor $\overline{g_1}_{\overline{\mathcal{H}}}$ in $\overline{\mathcal{H}}$ by

$$(\overline{g_1}_{\overline{\mathcal{H}}})_{p_1} = \overline{g_1}_{p_1 \overline{\mathcal{H}}(p_1)}.$$

Thus $(\overline{\mathcal{H}}, \overline{g_1}_{\overline{\mathcal{H}}})$ is a semi-Riemannian vector bundle over M_1 of some index $\overline{\nu}_1$. Hence it follows that the restriction $g_{1\mathcal{H}}$ of g_1 to $\mathcal{H}$ is of type $(\mu_1, \overline{\nu}_1, \eta_1)$ at every $p_1 \in M_1$, where $\mu_1 = \text{rank } L_1$ and η_1 is a certain integer. (See Remark 1.1.3).

Now since $\mathcal{V} = \mathcal{H}^\perp$, it follows from Proposition 1.1.2 that the restriction $g_{1\mathcal{V}}$ of g_1 to $\mathcal{V}$ is of type $(\mu_1, \nu_1 - \mu_1 - \overline{\nu}_1, n_1 - \nu_1 - \mu_1 - \eta_1)$ at every $p_1 \in M_1$. Thus fibers of f are degenerate submanifolds of constant type $(\mu_1, \nu_1 - \mu_1 - \overline{\nu}_1, n_1 - \nu_1 - \mu_1 - \eta_1)$, provided that rank $L_1 \neq 0$.

On the other hand, since f is a submersion, range $f_{*p_1} = T_{p_2}M_2$ at each $p_1 \in M_1$. Thus $L_2(p_2) = \{0\}$ and therefore, $(A_2(p_2), g_{2p_2 A_2(p_2)})$ is canonically isometric to $(\overline{A_2}(p_2), \overline{g_2}_{p_2 \overline{A_2}(p_2)})$ by $\pi_{A_2(p_2)}$ at each $p_2 \in f(M_1)$.

Now by identifying $(A_2(p_2), g_{2p_2 A_2(p_2)})$ and $(\overline{A_2}(p_2), \overline{g_2}_{p_2 \overline{A_2}(p_2)})$ at each $p_2 \in f(M_1)$, we define

$$\overline{A_2} = A_2 = \bigcup_{p_2 \in f(M_1)} A_2(p_2) = TM_2 \mid_{f(M_1)} .$$

Note that since submersions are open maps, $f(M_1)$ is an open submanifold of M_2 and hence, $\overline{A_2} = A_2$ is a vector bundle over $f(M_1)$. Thus we can extend $\overline{f_{*p_1}}$ defined at each $p_1 \in M_1$ to a linear map

$$\overline{f_*} : \overline{\mathcal{H}} \to \overline{A_2} = A_2$$

by

$$\overline{f_*}\overline{x} = \overline{f_{*p_1}}\overline{x} = f_{*p_1}x$$

where $\overline{x} \in \overline{\mathcal{H}}(p_1)$ and $x \in \mathcal{H}(p_1)$ with $\pi_{\mathcal{H}(p_1)}(x) = \overline{x}$.

Then note that, since $\overline{f_{*p_1}}$ is an (into) isometry at each $p_1 \in M_1$,

$$\overline{f_*} : (\overline{\mathcal{H}}, \overline{g_{1\overline{\mathcal{H}}}}) \to (A_2, g_{2f(M_1)})$$

is an (into) isometric linear map between semi-Riemannian vector bundles $(\overline{\mathcal{H}}, \overline{g_{1\overline{\mathcal{H}}}})$ and $(A_2, g_{2f(M_1)})$. Now let

$$\mathcal{B} = \text{range } \overline{f_*}.$$

Observe that, if $\overline{f_{*p_1}}(\overline{\mathcal{H}}(p_1)) = \overline{f_{*q_1}}(\overline{\mathcal{H}}(q_1))$ for all $p_1,\ q_1 \in M_1$ with $f(p_1) = f(q_1)$ then $\mathcal{B} = \text{range } \overline{f_*}$ is a subbundle of A_2. In this case, we call $\mathcal{B}$ an *$\overline{f_*}$-subbundle of A_2*. Note that, if $\mathcal{B}$ is an $\overline{f_*}$-subbundle of A_2 then $(\mathcal{B}, g_{2\mathcal{B}})$ is a semi-Riemannian subbundle of $(A_2, g_{2f(M_1)})$ and

$$\overline{f_*} : (\overline{\mathcal{H}}, \overline{g_{1\overline{\mathcal{H}}}}) \to (\mathcal{B}, g_{2\mathcal{B}})$$

is an isometric bundle map, where $g_{2\mathcal{B}}$ is the restriction of $g_{2f(M_1)}$ to $\mathcal{B}$.

Finally note that, if rank $L_1 = 0$ then $(\mathcal{V}, g_{1\mathcal{V}})$ is a semi-Riemannian subbundle of (TM_1, g_1) of some index ν_1' by Theorem 2.1.2 and hence, the fibers of f are semi-Riemannian submanifolds of index ν_1'. In this case, by identifying $(\mathcal{H}, g_{1\mathcal{H}})$ and $(\overline{\mathcal{H}}, \overline{g_{1\overline{\mathcal{H}}}})$, one can easily see that Definition 4.2.1 reduces to the usual definition of semi-Riemannian submersion (cf. [O, p. 212]). In the case of (M_1, g_1) and (M_2, g_2) are Riemannian manifolds, we call a semi-Riemannian submersion $f : (M_1, g_1) \to (M_2, g_2)$ a *Riemannian submersion.*

Proposition 4.2.1 *Let $f : (M_1, g_1) \to (M_2, g_2)$ be a semi-Riemannian submersion, where M_1 is connected and $\mathcal{B}$ is an $\overline{f_*}$-subbundle of A_2. If $A_1 = (\ker f_*) + (\ker f_*)^\perp$ is integrable then $\mathcal{B} = \text{range}\,\overline{f_*}$ is integrable and the integral manifolds of $\mathcal{B}$ are semi-Riemannian submanifolds of (M_2, g_2) of the same index.*

Proof. First note that $f_*(A_1) = \mathcal{B}$. Now let $X,\ Y \in \Gamma\mathcal{B}$ and let $\tilde{X},\ \tilde{Y} \in \Gamma A_1$ be lifts of X, Y to A_1, respectively, that is, $f_*\tilde{X} = X \circ f$ and $f_*\tilde{Y} = Y \circ f$. Then since $f_*([\tilde{X}, \tilde{Y}]) = [X, Y] \circ f$ and $[\tilde{X}, \tilde{Y}] \in \Gamma A_1$, it follows that $[X, Y] \in \Gamma\mathcal{B}$. That is, $\mathcal{B}$ is integrable. Thus by Remark 4.2.1, integral manifolds of $\mathcal{B}$ are semi-Riemannian submanifolds of (M_2, g_2) of the same index. □

Definition 4.2.2 A semi-Riemannian submersion $f : (M_1, g_1) \to (M_2, g_2)$ is called *regular* if $A_1 = (\ker f_*) + (\ker f_*)^\perp$ is a totally geodesic distribution, that is, $\overset{1}{\nabla}_X Y \in \Gamma A_1$ for every $X,\ Y \in \Gamma A_1$, where $\overset{1}{\nabla}$ is the Levi-Civita connection of (M_1, g_1).

Remark 4.2.2 Let $f : (M_1, g_1) \to (M_2, g_2)$ be a regular semi-Riemannian submersion, where M_1 is connected and $\mathcal{B}$ is an $\overline{f_*}$-subbundle of A_2. The first implication of the regularity of f is that A_1 is integrable, and hence $\mathcal{B}$ is integrable by Proposition 4.2.1. (Indeed, if X, $Y \in \Gamma A_1$ then, since $[X,Y] = \overset{1}{\nabla}_X Y - \overset{1}{\nabla}_Y X \in \Gamma A_1$, A_1 is integrable). The second implication of the regularity is on the fibers of f when they are degenerate submanifolds. That is, regularity of f guarantees that these degenerate submanifolds as fibers of f are stationary and irrotational. (See Proposition A.2.5). To be stationary and irrotational for a degenerate submanifold is an important property which enables one to relate the intrinsic structure of a degenerate submanifold to the geometry of the ambient semi-Riemannian manifold. (See Theorem A.2.2). Also we make use of regularity as a technical assumption which enable us to obtain O'Neill's formula for a semi-Riemannian submersion when fibers are degenerate submanifolds. Finally note that, if f is a semi-Riemannian submersion with rank $L_1 = 0$ then f is regular.

The semi-Riemannian nature of a semi-Riemannian submersion enables us to compare curvatures of manifolds. Next we state the O'Neill's formula for a semi-Riemannian submersion as such a comparison of curvatures of manifolds.

Let $f : (M_1, g_1) \to (M_2, g_2)$ be a semi-Riemannian submersion, where M_1 is connected and $\mathcal{B}$ is an $\overline{f_*}$-subbundle of A_2. Let $\mathcal{H}'$ be a complementary subbundle to L_1 in $\mathcal{H}$ and let $\mathcal{V}'$ be a complementary subbundle to L_1 in $\mathcal{V}$. Then note that

$$A_1 = \mathcal{H}' \oplus L_1 \oplus \mathcal{V}'$$

is a direct sum decomposition of A_1. Thus, if $U \in \Gamma A_1$ then U can be written uniquely as

$$U = U^{\perp} + U^0 + U^T,$$

where $U^{\perp}$, U^0 and U^T are the components of U in $\Gamma\mathcal{H}'$, ΓL_1 and $\Gamma\mathcal{V}'$, respectively. Also note that, if $g_{1\mathcal{H}'}$ is the restriction of g_1 to $\mathcal{H}'$ then $(\mathcal{H}', g_{1\mathcal{H}'})$ is a semi-Riemannian subbundle of (TM_1, g_1) and

$$f_* : (\mathcal{H}', g_{1\mathcal{H}'}) \to (\mathcal{B}, g_{2\mathcal{B}})$$

becomes an isometric bundle map. Thus, in particular, for every $X \in \Gamma\mathcal{B}$, there is a unique lift $\tilde{X}$ of X to $\mathcal{H}'$, that is, $\tilde{X} \in \Gamma\mathcal{H}'$ with $f_*\tilde{X} = X \circ f$.

Theorem 4.2.1 [K3] *Let $f : (M_1, g_1) \to (M_2, g_2)$ be a regular semi-Riemannian submersion, where M_1 is connected and $\mathcal{B}$ is an $\overline{f_*}$-subbundle of A_2. If $\tilde{X}$, $\tilde{Y} \in \Gamma\mathcal{H}'$ are the lifts of X, $Y \in \Gamma\mathcal{B}$ spanning nondegenerate planes at each point then*

$$K_{N_2}(\text{span}\{X,Y\}) \circ f = K_{M_1}(\text{span}\{\tilde{X},\tilde{Y}\}) + \tfrac{3}{4} g_1([\tilde{X},\tilde{Y}]^T,[\tilde{X},\tilde{Y}]^T)/Q(\tilde{X},\tilde{Y}),$$

where K_{N_2} is the sectional curvature of the integral manifolds (N_2, g_{2N_2}) of $\mathcal{B}$, K_{M_1} is the sectional curvature of (M_1, g_1) and $Q(\tilde{X},\tilde{Y}) = g_1(\tilde{X},\tilde{X})\, g_1(\tilde{Y},\tilde{Y}) - g_1(\tilde{X},\tilde{Y})^2$. (See Definition 2.3.3).

In Chapter 5, we will introduce semi-Riemannian transversal maps as generalizations of semi-Riemannian submersions. There we will state a similar theorem to Theorem 4.2.1 for semi-Riemannian transversal maps. (See Theorem 5.2.1). Then, since semi-Riemannian submersions are special cases of semi-Riemannian transversal maps, Theorem 4.2.1 will be a special case of the corresponding theorem for semi-Riemannian transversal maps. (Cf. Theorem 5.2.1).

Application of semi-Riemannian maps to immersions between semi-Riemannian manifolds does not enable us to obtain a generalization of isometric immersion as in the case of semi-Riemannian submersion. Recall that, an immersion $f : (M_1, g_1) \to (M_2, g_2)$ is called *isometric* if $g_1 = f^* g_2$. If one wants to generalize this definition by calling an immersion $f : (M_1, g_1) \to (M_2, g_2)$ isometric if f is a semi-Riemannian map, then this does not yield a generalization of the usual definition of isometric immersion. Indeed, first note that if $f : (M_1, g_1) \to (M_2, g_2)$ is an immersion then $(\mathcal{H}(p_1), g_{1p_1 \mathcal{H}(p_1)}) = (T_{p_1}M_1, g_{1p_1})$ is canonically isometric to $(\overline{\mathcal{H}}(p_1), \overline{g_1}_{p_1 \overline{\mathcal{H}}(p_1)})$ by the natural projection $\pi_{\mathcal{H}(p_1)}$ and dim $\overline{A}(p_2) =$ dim $A_2(p_2) -$ dim $L_2(p_2)$, where dim $A_2(p_2) =$ dim $\mathcal{H}(p_1)$. Thus, if f is semi-Riemannian then, since $\overline{f_{*p_1}} : (\overline{\mathcal{H}}(p_1), \overline{g_1}_{p_1 \overline{\mathcal{H}}(p_1)}) \to (\overline{A_2}(p_2), \overline{g_2}_{p_2 \overline{A_2}(p_2)})$ is an (into) isometry, dim $\overline{A_2}(p_2) \geq$ dim $\mathcal{H}(p_1)$. Hence, $L_2(p_2) = \{0\}$ and $(A_2(p_2), g_{2p_2 A_2(p_2)})$ is canonically isometric to $(\overline{A_2}(p_2), \overline{g_2}_{p_2 \overline{A_2}(p_2)})$ by the natural projection $\pi_{A_2(p_2)}$. Therefore $\overline{f_{*p_1}} : (\overline{\mathcal{H}}(p_1), \overline{g_1}_{p_1 \overline{\mathcal{H}}(p_1)}) \to (\overline{A_2}(p_2), \overline{g_2}_{p_2 \overline{A_2}(p_2)})$ is an isometry at each $p_1 \in M_1$, that is, $g_1 = f^* g_2$ as in the usual definition of an isometric immersion.

4.3 Riemannian Maps

Let $f : (M_1, g_1) \to (M_2, g_2)$ be a semi-Riemannian map. Note that, although the nondegenerate rank of f is constant on each connected component of M_1, the rank of f may change from point to point. Therefore, in general, we cannot use the methods of differential topology in further study of semi-Riemannian maps. Also other difficulties coming from the linear algebra of indefinite inner product spaces are apparent. However a natural special case smooths out all these difficulties, that is, assuming (M_1, g_1) and (M_2, g_2) are Riemannian manifolds.

Let $f : (M_1, g_1) \to (M_2, g_2)$ be a map between Riemannian manifolds (M_1, g_1) and (M_2, g_2). Also let $p_2 = f(p_1)$ for each $p_1 \in M_1$. Note that then, since $(\mathcal{H}(p_1),\, g_{1_{p_1 \mathcal{H}(p_1)}})$ and $(A_2(p_2),\, g_{2_{p_2 A_2(p_2)}})$ are positive definite inner product spaces, $(\mathcal{H}(p_1),\, g_{1_{p_1 \mathcal{H}(p_1)}})$ and $(A_2(p_2),\, g_{2_{p_2 A_2(p_2)}})$ are canonically isometric to $(\overline{\mathcal{H}}(p_1),\, \overline{g_1}_{p_1 \overline{\mathcal{H}}(p_1)})$ and $(\overline{A_2}(p_2),\, \overline{g_2}_{p_2 \overline{A_2}(p_2)})$ by the natural projections $\pi_{\mathcal{H}(p_1)}$ and $\pi_{A_2(p_2)}$, respectively. (See section 1.2). Hence by identifying $(\mathcal{H}(p_1),\, g_{1_{p_1 \mathcal{H}(p_1)}})$ and $(A_2(p_2),\, g_{2_{p_2 A_2(p_2)}})$ with $(\overline{\mathcal{H}}(p_1),\, \overline{g_1}_{p_1 \overline{\mathcal{H}}(p_1)})$ and $(\overline{A_2}(p_2),\, \overline{g_2}_{p_2 \overline{A_2}(p_2)})$, respectively, we can identify the quotient $\overline{f}_{*p_1} : \overline{\mathcal{H}}(p_1) \to \overline{A_2}(p_2)$ of f_{*p_1} by

$$f^h_{*\,p_1} : \mathcal{H} \to A_2$$

defined by

$$f^h_{*\,p_1} x = f_{*p_1} x,$$

where $x \in \mathcal{H}$. Here it is important to observe that $f^h_{*\,p_1}$ is an isomorphism and the nondegenerate rank of f at p_1 is equal to the rank of $f^h_{*\,p_1}$ which is equal to the rank of f at $p_1 \in M_1$. (See section 1.2).

Definition 4.3.1 Let (M_1, g_1) and (M_2, g_2) be Riemannian manifolds. Then a semi-Riemannian map $f : (M_1, g_1) \to (M_2, g_2)$ at $p_1 \in M_1$ is called a *Riemannian map at* p_1. In other words, a map $f : (M_1, g_1) \to (M_2, g_2)$ is called *Riemannian* at $p_1 \in M_1$ if $f^h_{*\,p_1} : (\mathcal{H}(p_1), g_{1_{p_1 \mathcal{H}(p_1)}}) \to (A_2(p_2), g_{2_{p_2 A_2(p_2)}})$ is an isometry. If $f : (M_1, g_1) \to (M_2, g_2)$ is Riemannian at each $p_1 \in M_1$ then f is called a *Riemannian map.*

Remark 4.3.1 Note that, $f : (M_1, g_1) \to (M_2, g_2)$ is Riemannian at $p_1 \in M_1$ is equivalent to $f^h_{*\,p_1}$ is Euclidean. (See Definition 1.2.6).

Remark 4.3.2 Note that if $f : (M_1, g_1) \to (M_2, g_2)$ is a Riemannian map then $\|f_*\|^2 = \operatorname{rank} f$ by Theorem 4.1.1. Hence the rank of a Riemannian map $f : (M_1, g_1) \to (M_2, g_2)$ is constant on each connected component of M_1.

Now let $f : (M_1, g_1) \to (M_2, g_2)$ be a map between Riemannian manifolds (M_1, g_1) and (M_2, g_2). Considering $f^h_{*\,p_1}$ at each $p_1 \in M_1$ as a linear transformation $f^h_{*\,p_1} : (\mathcal{H}(p_1), g_{1_{p_1 \mathcal{H}(p_1)}}) \to (A_2(p_2), g_{2_{p_2 A_2(p_2)}})$, we will denote the adjoint of $f^h_{*\,p_1}$ by ${}^*(f^h_{*\,p_1})$. Also let ${}^*f_{*p_1}$ be the adjoint of $f_{*p_1} : (T_{p_1}M_1, g_{1_{p_1}}) \to (T_{p_2}M_2,\, g_{2_{p_2}})$. Then by Proposition 1.2.1, the linear transformation

$$({}^*f_{*p_1})^h : A_2(p_2) \to \mathcal{H}(p_1)$$

defined by

$$({}^*f_{*p_1})^h y = {}^*f_{*p_1} y,$$

where $y \in A_2$, is an isomorphism. (See section 1.2). Finally as in section 1.2, define linear transformations

$$P_{p_1} = {}^*f_{*p_1} \circ f_{*p_1} : (T_{p_1}M_1, g_{1p_1}) \to (T_{p_1}M_1, g_{1p_1})$$

and

$$Q_{p_1} = f_{*p_1} \circ {}^*f_{*p_1} : (T_{p_2}M_2, g_{2p_2}) \to (T_{p_2}M_2, g_{2p_2})$$

at each $p_1 \in M_1$. Also let $P^h_{p_1} : \mathcal{H}(p_1) \to \mathcal{H}(p_1)$ and $Q^h_{p_1} : A_2(p_2) \to A_2(p_2)$ be the isomorphisms obtained by the restrictions of P_{p_1} and Q_{p_1} to $\mathcal{H}(p_1)$ and $A_2(p_2)$, respectively, as in section 1.2. Then by Proposition 1.2.4, we have the following characterization of Riemannian maps.

Proposition 4.3.1 [F1] *Let $f : (M_1, g_1) \to (M_2, g_2)$ be a map between Riemannian manifolds (M_1, g_1) and (M_2, g_2). Then the following are equivalent:*

(1) *f is Riemannian at $p_1 \in M_1$.*

(2) *P_{p_1} is a projection, that is, $P_{p_1} \circ P_{p_1} = P_{p_1}$.*

(3) *Q_{p_1} is a projection, that is, $Q_{p_1} \circ Q_{p_1} = Q_{p_1}$.*

(4) *$P^h_{p_1} = id_{\mathcal{H}(p_1)}$, that is, $P^h_{p_1}$ is the identity on $\mathcal{H}(p_1)$.*

(5) *$Q^h_{p_1} = id_{A_2(p_2)}$, that is, $Q^h_{p_1}$ is the identity on $A_2(p_2)$.*

(6) *$(f^h_{*p_1})^{-1} = ({}^*f_{*p_1})^h = {}^*(f^h_{*p_1})$.*

Note that such a characterization is not available for semi-Riemannian maps because of the possibility of degenerate kernels and ranges.

Next we investigate the global properties of Riemannian maps.

We recall from differential topology that, a map $f : M_1 \to M_2$ is called *subimmersion at* $p_1 \in M_1$ if there is a neighborhood U of p_1, a manifold P, a submersion $s : U \to P$, and an immersion $j : P \to M_2$ such that $f\,|_U = f_U = j \circ s$. A map $f : M_1 \to M_2$ is called a *subimmersion* if it is subimmersion at each $p_1 \in M_1$. It is well-known that a map $f : M_1 \to M_2$ is a subimmersion if and only if the rank of f is locally constant, and hence constant on each connected component of M_1. (Cf. [AMR, p. 205]). Thus by Remark 4.3.2, we obtain the following proposition.

Proposition 4.3.2 *A Riemannian map $f : (M_1, g_1) \to (M_2, g_2)$ between Riemannian manifolds (M_1, g_1) and (M_2, g_2) is a subimmersion.*

Furthermore, being Riemannian for f makes s and j in the definition of subimmersion a Riemannian submersion and an isometric immersion, respectively. To show this, first we make the following observation.

Lemma 4.3.1 *A map $f : M_1 \to (M_2, g_2)$ between a manifold M_1 and a Riemannian manifold (M_2, g_2) is an immersion if and only if $g_1 = f^*g_2$ is a positive definite metric tensor on M_1, that is, (M_1, g_1) is a Riemannian manifold. In this case, $f : (M_1, g_1) \to (M_2, g_2)$ is an isometric immersion.*

Proof. Clearly g_1 is a symmetric $(0, 2)$-tensor field on M_1. Thus for $0 \neq x \in T_{p_1}M_1$, g_1 is a positive definite metric tensor if and only if $g_1(x, x) > 0$ if and only if $f_{*p_1}x \neq 0$ (since g_2 is positive definite) if and only if ker $f_{*p_1} = \{0\}$ if and only if f is an immersion. □

Proposition 4.3.3 [F1] *Let $f : (M_1, g_1) \to (M_2, g_2)$ be a Riemannian map between Riemannian manifolds (M_1, g_1) and (M_2, g_2). Then by Proposition 4.3.2, since f is a subimmersion, let U, P, s and j be as in the definition of subimmersion, so that $f|_U = f_U = j \circ s$. Let g_{1U} denote the restriction of g_1 to U and let $g_P = j^*g_2$. Then (U, g_{1U}) and (P, g_P) are Riemannian manifolds, the submersion $s : (U, g_{1U}) \to (P, g_P)$ is a Riemannian submersion, and the immersion $j : (P, g_P) \to (M_2, g_2)$ is an isometric immersion.*

Proof. First note that since U is open in M_1, (U, g_{1U}) is a Riemannian manifold. Also it follows from Lemma 4.3.1 that (P, g_P) is a Riemannian manifold and $j : (P, g_P) \to (M_2, g_2)$ is an isometric immersion. For each $p_1 \in U$, define

$$s^h_{*p_1} : \mathcal{H}(p_1) \to T_{s(p_1)}P \quad \text{by} \quad s^h_{*p_1}x = s_{*p_1}x,$$

and define

$$j^h_{*\,s(p_1)} : T_{s(p_1)}P \to A_2(p_2) \quad \text{by} \quad j^h_{*\,s(p_1)}z = j_{*s(p_1)}z.$$

Then

$$f^h_{*\,p_1} = (j_{*s(p_1)} \circ s_{*p_1})^h = j^h_{*\,s(p_1)} \circ s^h_{*p_1} : \mathcal{H}(p_1) \to A_2(p_2).$$

Since $f : (M_1, g_1) \to (M_2, g_2)$ is Riemannian,

$$f^h_{*\,p_1} : (\mathcal{H}(p_1), g_{1p_1\mathcal{H}(p_1)}) \to (A_2(p_2), g_{2p_2A_2(p_2)})$$

is an isometry and since $j : (P, g_P) \to (M_2, g_2)$ is an isometric immersion,

$$j^h_{*\,s(p_1)} : (T_{s(p_1)}P, g_{P\,s(p_1)}) \to (A_2(p_2), g_{2p_2A_2(p_2)})$$

is also an isometry. Since

$$f^h_{*\,p_1} = j^h_{*\,s(p_1)} \circ s^h_{*p_1},$$

and the left hand side and the first factor of the right hand side are isometries, it then follows that

$$s^h_{*p_1} : (\mathcal{H}(p_1), g_{1p_1\mathcal{H}(p_1)}) \to (T_{s(p_1)}P, g_{P_{s(p_1)}})$$

is an isometry. Hence $s : (U, g_{1U}) \to (P, g_P)$ is a Riemannian submersion. □

Notice that, by the above proposition, a Riemannian map between Riemannian manifolds may be considered to be the appropriate Riemannian generalization of the notion of subimmersion that occurs in the theory of maps between manifolds.

By using the methods of differential topology, we can also obtain the following global results for Riemannian maps.

Theorem 4.3.1 [F1] *Let $f : (M_1, g_1) \to (M_2, g_2)$ be a Riemannian map between Riemannian manifolds (M_1, g_1) and (M_2, g_2), where $\dim M_2 = n_2 \geq 1$.*

(1) *If f is injective then f is an isometric immersion.*

(2) *If f is surjective and M_1 is connected then f is a surjective Riemannian submersion.*

(3) *More generally, if f is open or closed onto its image then $f(M_1)$ is a submanifold of N. Let $g_{2f(M_1)}$ be the restriction of g_2 on $f(M_1)$. Then*

$$f : (M_1, g_1) \to (f(M_1), g_{2f(M_1)})$$

(also denoted by f) is a surjective Riemannian submersion.

(4) *If f is a bijection and M_1 is connected then f is an isometry.*

Proof. First note that f is a subimmersion by Proposition 4.3.2.

(1): Since f is a subimmersion, for each $p_1 \in M_1$, $f^{-1}(f(p_1))$ is a closed submanifold of M_1 with tangent space $T_{p_1}(f^{-1}(f(p_1))) = \ker f_{*p_1}$. (Cf. [AMR, p. 205]). Since f is injective, $f^{-1}(f(p_1)) = \{p_1\}$ so $\ker f_{*p_1} = \{0\}$. Thus f is an immersion and hence an injective immersion. Then since f is a Riemannian map, f is an isometric immersion.

(2): Let $M_1^{reg} = \{p_1 \in M_1 \mid f_{*p_1} \text{ is surjective}\}$ denote the regular points of f. Now since f is a subimmersion and M_1 is connected, rank of f is constant on M_1.

Thus either $M_1^{reg} = M_1$ or $M_1^{reg} = \emptyset$. But since f is surjective, it follows from Sard's theorem (see [BJ, p. 56]) that $M_1^{reg} \neq \emptyset$. Hence f is a surjective submersion. Then since f is a Riemannian map, f is a surjective Riemannian submersion.

(3): Since f is an open or closed subimmersion onto its image, $f(M_1)$ is a submanifold of M_2 (see [AMR, p. 206]), and then $f : M_1 \to f(M_1)$ (also denoted by f) is a surjective subimmersion. Then by the same argument in (2), $f : (M_1, g_1) \to (f(M_1), g_{2f(M_1)})$ is a surjective Riemannian submersion.

(4): Since f is a subimmersion, if f is bijective and M_1 is connected then f is an injective immersion and a surjective submersion, hence a bijective local diffeomorphism which is a diffeomorphism. Thus, since f is Riemannian, it is an isometry. □

Remark 4.3.3 Note that Theorem 4.3.1–(3) also indicates that Riemannian maps compare geometric properties of the manifolds in their domains and ranges. (See Theorem 4.2.1).

An application of Theorem 4.3.1 can be made in weakening the hypothesis of a famous theorem of Hermann. Recalling Hermann's Theorem:

Theorem 4.3.2 [He] *Let $f : (M_1, g_1) \to (M_2, g_2)$ be a surjective Riemannian submersion between Riemannian manifolds (M_1, g_1) and (M_2, g_2). Assume that M_1 is connected and (M_1, g_1) is complete. Then f is a locally trivial fiber bundle over M_2.*

Now as a consequence of Theorem 4.3.1–(2), a surjective Riemannian map on a connected manifold is automatically a surjective Riemannian submersion. Thus we have the following strengthening of Hermann's Theorem:

Theorem 4.3.3 [F1] *Let $f : (M_1, g_1) \to (M_2, g_2)$ be a surjective Riemannian map between Riemannian manifolds (M_1, g_1) and (M_2, g_2). Assume that M_1 is connected and (M_1, g_1) is complete. Then f is a surjective Riemannian submersion and f is a locally trivial fiber bundle over M_2.*

4.4 Affine Riemannian Maps

A semi-Riemannian map is a map that is as "isometric as it can be". Consequently, as we have seen in Theorem 4.2.1, a semi-Riemannian map compares geometric properties of the manifolds in its domain and range, subject to the limitations imposed upon it as a mapping. (Also see Remark 4.3.3). In particular, if we impose the condition of being affine on a Riemannian map then it becomes a "partial isometry". We consider an affine Riemannian map as a "partial isometry" because of the following observation: Let $f : (M_1, g_1) \to (M_2, g_2)$ be an affine Riemannian map between Riemannian manifolds (M_1, g_1) and (M_2, g_2). First note that since f is affine, by Proposition 3.2.2–(a), $f_* \overset{1}{R}(X,Y)Z = \overset{2}{R}(f_*X, f_*Y)f_*Z$, for all X, Y, $Z \in \Gamma TM_1$. Therefore, if x, $y \in \mathcal{H}(p_1)$ are orthonormal vectors then span$\{x, y\}$ and span$\{f_{*p_1}x, f_{*p_1}y\}$ are planes in $\mathcal{H}(p_1)$ and $A_2(p_2)$, respectively, and

$$\begin{aligned} K_1(\text{span}\,\{x,y\}) &= g_1(\overset{1}{R}(x,y)y, x) \\ &= g_1((\overset{1}{R}(x,y)y)^{\perp}, x) \\ &= g_2(f_{*p_1}(\overset{1}{R}(x,y)y)^{\perp}, f_{*p_1}x) \\ &= g_2(f_{*p_1}\overset{1}{R}(x,y)y, f_{*p_1}x) \\ &= g_2(\overset{2}{R}(f_{*p_1}x, f_{*p_1}y)f_{*p_1}y, f_{*p_1}x) \\ &= K_2(\text{span}\,\{f_{*p_1}x, f_{*p_1}y\}), \end{aligned}$$

where K_1 and K_2 denote the sectional curvatures of (M_1, g_1) and (M_2, g_2), respectively, and $(\overset{1}{R}(x,y)y)^{\perp}$ is the component of $\overset{1}{R}(x,y)y$ in $\mathcal{H}(p_1)$. Thus, since an affine Riemannian map preserves sectional curvatures in the above sense, it may be considered as a "partial isometry". Therefore, in this section we investigate some sufficient conditions for a Riemannian map to be affine. More explicitly, we concentrate on curvature conditions under which a Riemannian map $f : (M_1, g_1) \to (M_2, g_2)$ with div $\tau(f) \geq 0$ is affine.

Remark 4.4.1 It is important to note that, if $f : (M_1, g_1) \to (M_2, g_2)$ is a Riemannian map between Riemannian manifolds (M_1, g_1) and (M_2, g_2) with div $\tau(f) \geq 0$, where (M_1, g_1) is orientable and compact, then by Proposition 3.3.1, f is a harmonic Riemannian map. But in general, a Riemannian map may not be harmonic. For example $f : (\mathbf{R}, dt \otimes dt) \to (\mathbf{R}^2, dx^1 \otimes dx^1 + dx^2 \otimes dx^2)$ given by $f(t) = (\cos t, \sin t)$ is a Riemannian map but not a harmonic map. (See Proposition 3.2.4). In fact, div $\tau(f) = -1$.

Let (M_1, g_1) and (M_2, g_2) be Riemannian manifolds of dimensions n_1 and n_2, respectively. Let r_1 and S_1 denote the Ricci and scalar curvatures of (M_1, g_1), respectively, and, let K_2 and S_2 denote the sectional and scalar curvatures of (M_2, g_2), respectively. (See Section 2.3). By $r_1 \geq A$, we mean $r_1(v) \geq A$ for every $0 \neq v \in TM_1$, and by $K_2 \leq B$, we mean $K_2(P) \leq B$ for every plane in TM_2.

Lemma 4.4.1 *Let $f : (M_1, g_1) \to (M_2, g_2)$ be a Riemannian map between Riemannian manifolds (M_1, g_1) and (M_2, g_2). Then at each $p_1 \in M_1$,*

$$\begin{aligned} 0 &= \sum_{i,j=1}^{n_1} g_2((\nabla f_*)(x_i, x_j), (\nabla f_*)(x_i, x_j)) \\ &\quad - \sum_{i,j=m_1+1}^{n_1} g_2(\overset{2}{R}(f_* x_i, f_* x_j) f_* x_j, f_* x_i) \\ &\quad + \sum_{i=m_1+1}^{n_1} r_1(x_i) \\ &\quad +(\operatorname{div} \tau(f))(p_1) \end{aligned}$$

where $\{x_1, \ldots, x_{m_1}, x_{m_1+1}, \ldots, x_{n_1}\}$ *is an orthonormal basis for* $(T_{p_1}M_1, g_{1p_1})$ *with* $\{x_1, \ldots, x_{m_1}\}$ *is a basis for* $\mathcal{V}(p_1) = \ker f_{*p_1}$.

Proof. Since $P^h_{p_1} = ({}^*f_{*p_1} \circ f_{*p_1}) |_{\mathcal{H}(p_1)}: \mathcal{H}(p_1) \to \mathcal{H}(p_1)$ is the identity by Proposition 4.3.1 and $\|f_*\|^2 = \operatorname{rank} f$ is constant on each connected component of M_1 (cf. Remark 4.3.2), for such a basis $\{x_1, \ldots, x_{m_1}, x_{m_1+1}, \ldots, x_{n_1}\}$ for $(T_{p_1}M_1, g_{1p_1})$, the Bochner identity (Lemma 3.4.1) becomes

$$\begin{aligned} 0 &= \sum_{i,j=1}^{n_1} g_2((\nabla f_*)(x_i, x_j), (\nabla f_*)(x_i, x_j)) \\ &\quad - \sum_{i,j=m_1+1}^{n_1} g_2(\overset{2}{R}(f_* x_i, f_* x_j) f_* x_j, f_* x_i) \\ &\quad + \sum_{\substack{i=m_1+1 \\ j=1}}^{n_1} g_1(x_i, \overset{1}{R}(x_i, x_j) x_j) \\ &\quad +(\operatorname{div} \tau(f))(p_1) \\ &= \sum_{i,j=1}^{n_1} g_2((\nabla f_*)(x_i, x_j), (\nabla f_*)(x_i, x_j)) \\ &\quad - \sum_{i,j=m_1+1}^{n_1} g_2(\overset{2}{R}(f_* x_i, f_* x_j) f_* x_j, f_* x_i) \\ &\quad + \sum_{i=m_1+1}^{n_1} r_1(x_i) \\ &\quad +(\operatorname{div} \tau(f))(p_1). \end{aligned}$$

□

Theorem 4.4.1 [GK2] *Let (M_1, g_1) be a connected Riemannian manifold with $r_1 \geq A$ and let (M_2, g_2) be a Riemannian manifold with $K_2 \leq B$. If $f : (M_1, g_1) \to (M_2, g_2)$ is a Riemannian map with* rank $f \geq 2$, $A \geq (\text{rank } f - 1)B$ *and* div $\tau(f) \geq 0$, *then f is affine.*

Proof. Let $\{x_1,\ldots,x_{m_1},x_{m_1+1},\ldots,x_{n_1}\}$ be an orthonormal basis for $(T_{p_1}M_1, g_{1p_1})$ with $\{x_1,\ldots,x_{m_1}\}$ is a basis for $\mathcal{V}(p_1) = \ker f_{*p_1}$. Then,

$$-\sum_{i,j=m_1+1}^{n_1} g_2(\overset{2}{R}(f_*x_i, f_*x_j)f_*x_j, f_*x_i) \geq -(\text{rank } f)(\text{rank } f - 1)B$$

and

$$\sum_{i=m_1+1}^{n_1} r_1(x_i) \geq (\text{rank } f)A.$$

Hence we obtain from Lemma 4.4.1 that,

$$\begin{aligned} 0 \geq & \sum_{i,j=1}^{n_1} g_2((\nabla f_*)(x_i, x_j), (\nabla f_*)(x_i, x_j)) \\ & -(\text{rank } f)(\text{rank } f - 1)B + (\text{rank } f)A + (\text{div } \tau(f))(p_1) \\ \geq & \sum_{i,j=1}^{n_1} g_2((\nabla f_*)(x_i, x_j), (\nabla f_*)(x_i, x_j)) + (\text{div } \tau(f))(p_1) \end{aligned}$$

at each $p_1 \in M_1$. Thus since div $\tau(f) \geq 0$,

$$g_2((\nabla f_*)(x_i, x_j), (\nabla f_*)(x_i, x_j)) = 0$$

for each $1 \leq i, j \leq n_1$, and it follows that $(\nabla f_*)_{p_1} = 0$ at each $p_1 \in M_1$. Hence f is an affine map. □

Next we state a version of Theorem 4.4.1 for Riemannian maps of rank 1.

Theorem 4.4.2 [GK2] *Let (M_1, g_1) be a connected Riemannian manifold with $r_1 \geq 0$ and let (M_2, g_2) be a Riemannian manifold. If $f : (M_1, g_1) \to (M_2, g_2)$ is a Riemannian map with* rank $f = 1$ *and* div $\tau(f) \geq 0$, *then f is affine.*

Proof. Let $\{x_1,\ldots,x_{n_1-1},x_{n_1}\}$ be an orthonormal basis for $(T_{p_1}M_1, g_{1p_1})$ with $\{x_1,\ldots,x_{n_1-1}\}$ is a basis for $\mathcal{V}(p_1) = \ker f_{*p_1}$. Note that then,

$$g_2(\overset{2}{R}(f_*x_i, f_*x_j)f_*x_j, f_*x_i) = 0$$

for $i, j = n_1$. Hence we obtain from Lemma 4.4.1 that

$$0 \geq \sum_{i,j=1}^{n_1} g_2((\nabla f_*)(x_i, x_j), (\nabla f_*)(x_i, x_j)) + (\mathrm{div}\ \tau(f))(p_1)$$

at each $p_1 \in M_1$. Thus since div $\tau(f) \geq 0$, it follows as in the proof of Theorem 4.4.1 that f is an affine map. □

Note that isometric immersions and Riemannian submersions between Riemannian manifolds are also Riemannian maps. Hence we have the following special cases of Theorem 4.4.1 for such maps.

Theorem 4.4.3 [GK2] *Let (M_1, g_1) be a connected Riemannian manifold with $n_1 \geq 2$ and $S_1 \geq A$, and let (M_2, g_2) be a Riemannian manifold with $K_2 \leq B$ such that $A \geq n_1(n_1 - 1)B$. If $f : (M_1, g_1) \to (M_2, g_2)$ is an isometric immersion with* div $\tau(f) \geq 0$, *then f is affine.*

Proof. Let $\{x_1, \ldots, x_{n_1}\}$ be an orthonormal basis for $(\mathcal{H}(p_1), g_{1_{p_1 \mathcal{H}(p_1)}}) = (T_{p_1}M_1, g_{1_{p_1}})$. Then since f is an immersion,

$$\begin{aligned} -\sum_{i,j=1}^{n_1} g_2(\overset{2}{R}(f_*x_i, f_*x_j)f_*x_j, f_*x_i) &\geq -(\mathrm{rank} f)(\mathrm{rank} f - 1)B \\ &= -n_1(n_1 - 1)B \end{aligned}$$

and

$$\sum_{i=1}^{n_1} r_1(x_i) = S_1(p_1).$$

Hence we obtain from Lemma 4.4.1 that,

$$\begin{aligned} 0 &\geq \sum_{i,j=1}^{n_1} g_2((\nabla f_*)(x_i, x_j), (\nabla f_*)(x_i, x_j)) \\ &\quad -n_1(n_1 - 1)B + A + (\mathrm{div}\ \tau(f))(p_1) \\ &\geq \sum_{i,j=1}^{n_1} g_2((\nabla f_*)(x_i, x_j), (\nabla f_*)(x_i, x_j)) + (\mathrm{div}\ \tau(f))(p_1) \end{aligned}$$

at each $p_1 \in M_1$. Thus since div $\tau(f) \geq 0$, it follows as in the proof of Theorem 4.4.1 that f is an affine map. □

Note here that Theorem 4.4.2 also serves as a counterpart to Theorem 4.4.3 for rank 1 isometric immersions.

Theorem 4.4.4 [GK2] *Let (M_1, g_1) be a connected Riemannian manifold with $r_1 \geq A$ and let (M_2, g_2) be a Riemannian manifold with $S_2 \leq B$ such that $n_2 A \geq B$. If $f : (M_1, g_1) \to (M_2, g_2)$ is a Riemannian submersion with* div $\tau(f) \geq 0$, *then f is affine.*

Proof. Let $\{x_1, \ldots, x_{m_1}, x_{m_1+1}, \ldots, x_{n_1}\}$ be an orthonormal basis for $(T_{p_1} M_1, g_{1p_1})$ with $\{x_1, \ldots, x_{m_1}\}$ is a basis for $\mathcal{V}(p_1) = \ker f_{*p_1}$. Then

$$
\begin{aligned}
& -\sum_{i,j=m_1+1}^{n_1} g_2(\overset{2}{R}(f_* x_i, f_* x_j) f_* x_j, f_* x_i) \\
& \qquad = -2 \sum_{m_1+1 \leq i < j \leq n_1}^{n_1} K_2(\text{span}\, \{f_* x_i, f_* x_j\}) \\
& \qquad = -S_2(p_2)
\end{aligned}
$$

and

$$
\sum_{i=m_1+1}^{n_1} r_1(x_i) \geq (\text{rank } f) A = n_2 A.
$$

Hence we obtain from Lemma 4.4.1 that

$$
\begin{aligned}
0 \;\geq\; & \sum_{i,j=1}^{n_1} g_2((\nabla f_*)(x_i, x_j), (\nabla f_*)(x_i, x_j)) \\
& -B + n_2 A + (\text{div } \tau(f))(p_1) \\
\geq\; & \sum_{i,j=1}^{n_1} g_2((\nabla f_*)(x_i, x_j), (\nabla f_*)(x_i, x_j)) + (\text{div } \tau(f))(p_1)
\end{aligned}
$$

at each $p_1 \in M_1$. Thus since div $\tau(f) \geq 0$, it follows as in the proof of Theorem 4.4.1 that f is an affine map. □

Affine Riemannian submersions have special importance in obtaining splitting theorems for Riemannian manifolds. Later in chapter 7, we will obtain a splitting theorem for Riemannian manifolds by employing Theorem 4.4.4. (See Theorem 7.2.5).

Chapter 5

Semi-Riemannian Transversal Maps

In this chapter, we generalize the notion of a semi-Riemannian map to the notion of a semi-Riemannian map with respect to a semi-Riemannian foliation on the target manifold. If this semi-Riemannian foliation is taken to be the points of the target manifold, then the definition of a semi-Riemannian map with respect to such a foliation reduces to the definition of a semi-Riemannian map.

In section 5.1, we define semi-Riemannian maps with respect to a semi-Riemannian foliation as a generalization of semi-Riemannian maps. In particular, we show that a semi-Riemannian map with respect to a semi-Riemannian foliation satisfies the transversal generalized eikonal equation. In section 5.2, we define semi-Riemannian transversal maps to a semi-Riemannian foliation which are generalizations of semi-Riemannian submersions. Also we obtain a fundamental equation of a regular normal semi-Riemannian transversal map to a semi-Riemannian foliation. A special case of this fundamental equation will provide us a proof of O'Neill's formula for a semi-Riemannian submersion. (See Theorem 4.2.1).

Throughout this chapter, let (M_1, g_1) and (M_2, g_2) be semi-Riemannian manifolds of indices ν_1 and ν_2, and dimensions n_1 and n_2, respectively. Also by $\overset{1}{\nabla}$ and $\overset{2}{\nabla}$, we will denote the Levi-Civita connections of (M_1, g_1) and (M_2, g_2), respectively. As usual, let $\overset{1}{R}$ and $\overset{2}{R}$ denote the curvature tensors of (M_1, g_1) and (M_2, g_2), respectively.

5.1 The Transversal Generalized Eikonal Equation and Semi-Riemannian Maps With Respect To a Semi-Riemannian Foliation

Let (M_2, g_2) be a semi-Riemannian manifold and let $\mathcal{F} = \{S_\alpha\}_{\alpha \in A}$ be a foliation on M_2. Let g_{2S_α} be the restriction of g_2 to S_α for each $\alpha \in A$. If every $(S_\alpha, g_{2S_\alpha})$ for $\alpha \in A$ is a semi-Riemannian manifold of the same index ν', then we call $(\mathcal{F}, g_{2\mathcal{F}})$ a *semi-Riemannian foliation (of index ν') on* (M_2, g_2). Also, if $(\mathcal{F}, g_{2\mathcal{F}})$ is a semi-Riemannian foliation on (M_2, g_2) then define the *tangent bundle of* $(\mathcal{F}, g_{2\mathcal{F}})$ by

$$T(M_2, \mathcal{F}) = \bigcup_{\alpha \in A} \bigcup_{p_2 \in S_\alpha} T_{p_2} S_\alpha$$

and define the *orthogonal bundle of* $(\mathcal{F}, g_{2\mathcal{F}})$ by

$$T(M_2, \mathcal{F})^\perp = \bigcup_{\alpha \in A} \bigcup_{p_2 \in S_\alpha} T_{p_2} S_\alpha^\perp.$$

where $T_{p_2} S_\alpha^\perp$ is the orthogonal space to $T_{p_2} S_\alpha$. Note here that, if $g_{2T(M_2,\mathcal{F})}$ and $g_{2T(M_2,\mathcal{F})^\perp}$ are the restrictions of g_2 to $T(M_2, \mathcal{F})$ and $T(M_2, \mathcal{F})^\perp$, respectively, then $(T(M_2, \mathcal{F}), g_{2T(M_2,\mathcal{F})})$ and $(T(M_2, \mathcal{F})^\perp, g_{2T(M_2,\mathcal{F})^\perp})$ are semi-Riemannian subbundles of (TM_2, g_2).

Now let $(\mathcal{F}, g_{2\mathcal{F}})$ be a semi-Riemannian foliation on (M_2, g_2) and let $f : (M_1, g_1) \to (M_2, g_2)$ be a map. Also let $p_2 = f(p_1)$ for each $p_1 \in M_1$. Then, at each $p_1 \in M_1$, define a linear transformation

$$f_{*p_1}^\perp : (T_{p_1} M_1, g_{1p_1}) \to (T_{p_2}(M_2, \mathcal{F})^\perp, g_{2p_2}{}_{T_{p_2}(M_2,\mathcal{F})^\perp})$$

by

$$f_{*p_1}^\perp x = (f_{*p_1} x)^\perp,$$

where $(f_{*p_1} x)^\perp$ is the component of $f_{*p_1} x$ in $T_{p_2}(M_2, \mathcal{F})^\perp$.

As in section 1.2, set

$$\mathcal{V}^{\mathcal{F}}(p_1) = \ker f_{*p_1}^{\perp}$$

$$\mathcal{H}^{\mathcal{F}}(p_1) = (\ker f_{*p_1}^{\perp})^{\perp} = \mathcal{V}^{\mathcal{F}}(p_1)^{\perp}$$

$$L_1^{\mathcal{F}}(p_1) = (\ker f_{*p_1}^{\perp}) \cap (\ker f_{*p_1}^{\perp})^{\perp} = L_{g_1 p_1 \mathcal{V}^{\mathcal{F}}(p_1)} = L_{g_1 p_1 \mathcal{H}^{\mathcal{F}}(p_1)}$$

$$A_1^{\mathcal{F}}(p_1) = (\ker f_{*p_1}^{\perp}) + (\ker f_{*p_1}^{\perp})^{\perp} = L_1^{\mathcal{F}}(p_1)^{\perp}$$

$$A_2^{\mathcal{F}}(p_2) = \text{range } f_{*p_1}^{\perp}$$

$$L_2^{\mathcal{F}}(p_2) = (\text{range } f_{*p_1}^{\perp}) \cap (\text{range } f_{*p_1}^{\perp})^{\perp} = L_{g_2 p_2 A_2^{\mathcal{F}}(p_2)}.$$

Also let $(\overline{\mathcal{H}}^{\mathcal{F}}(p_1), \overline{g_1}_{p_1 \overline{\mathcal{H}}^{\mathcal{F}}(p_1)})$ and $(\overline{A_2}^{\mathcal{F}}(p_2), \overline{g_2}_{p_2 \overline{A_2}^{\mathcal{F}}(p_2)})$ be the quotient inner product spaces introduced in section 1.2 and let

$$\overline{f_{*p_1}}^{\perp} : (\overline{\mathcal{H}}^{\mathcal{F}}(p_1), \overline{g_1}_{p_1 \overline{\mathcal{H}}^{\mathcal{F}}(p_1)}) \to (\overline{A_2}^{\mathcal{F}}(p_2), \overline{g_2}_{p_2 \overline{A_2}^{\mathcal{F}}(p_2)})$$

be the quotient of $f_{*p_1}^{\perp}$. (Cf. section 1.2).

Here it is worth emphasizing that $\overline{f_{*p_1}}^{\perp}$ may be neither injective or surjective. In fact, rank $\overline{f_{*p_1}}^{\perp} \leq$ rank $f_{*p_1}^{\perp}$ − dim $L_1^{\mathcal{F}}(p_1)$. (See Remark 1.2.1).

Remark 5.1.1 It is important to observe that, if $(\mathcal{F}, g_{2\mathcal{F}})$ is the discrete foliation on (M_2, g_2), that is, if the leaves of $\mathcal{F}$ are the points of M_2, then $T(M_2, \mathcal{F})^{\perp} = TM_2$. Thus in this case, $f_{*p_1}^{\perp} = f_{*p_1}$ and hence, the definition of $\overline{f_{*p_1}}^{\perp}$ reduces to the definition of $\overline{f_{*p_1}}$ in section 4.1.

Definition 5.1.1 Let $f : (M_1, g_1) \to (M_2, g_2)$ be a map and $(\mathcal{F}, g_{2\mathcal{F}})$ be a semi-Riemannian foliation on (M_2, g_2). The *nondegenerate transversal rank of f with respect to $(\mathcal{F}, g_{2\mathcal{F}})$* at $p_1 \in M_1$ is defined to be the rank of $\overline{f_{*p_1}}^{\perp}$.

Definition 5.1.2 Let $f : (M_1, g_1) \to (M_2, g_2)$ be a map and $(\mathcal{F}, g_{2\mathcal{F}})$ be a semi-Riemannian foliation on (M_2, g_2). Then f is called *weakly semi-Riemannian with respect to $(\mathcal{F}, g_{2\mathcal{F}})$* at $p_1 \in M_1$ if $\|f_{*p_1}^{\perp}\|^2 =$ rank $\overline{f_{*p_1}}^{\perp}$ and, f is called *weakly semi-Riemannian with respect to $(\mathcal{F}, g_{2\mathcal{F}})$* if f is weakly semi-Riemannian with respect to $(\mathcal{F}, g_{2\mathcal{F}})$ at each $p_1 \in M_1$.

Remark 5.1.2 Note that, f is weakly semi-Riemannian with respect to a semi-Riemannian foliation $(\mathcal{F}, g_{2\mathcal{F}})$ at $p_1 \in M_1$ is equivalent to $f_{*p_1}^{\perp}$ is weakly semi-Euclidean. (See Definition 1.2.5). Also, if $(\mathcal{F}, g_{2\mathcal{F}})$ is the discrete foliation on (M_2, g_2) then the definition of weakly semi-Riemannian map with respect to $(\mathcal{F}, g_{2\mathcal{F}})$ reduces to the definition of a weakly semi-Riemannian map. (See Remark 5.1.1).

Remark 5.1.3 Let $f : (M_1, g_1) \to (M_2, g_2)$ be weakly semi-Riemannian with respect to a semi-Riemannian foliation $(\mathcal{F}, g_{2\mathcal{F}})$ on (M_2, g_2). Then note that since, the map

$$\|f_*{}^{\perp}\|^2 : M_1 \to \mathbf{R}$$

defined by

$$\|f_*{}^{\perp}\|^2(p_1) = \|f_{*p_1}^{\perp}\|^2$$

is a continuous function, the nondegenerate transversal rank of f with respect to $(\mathcal{F}, g_{2\mathcal{F}})$ is constant on each connected component of M_1, that is, rank $\overline{f_{*p_1}^{\perp}}$ does not change on each connected component of M_1.

Definition 5.1.3 Let (M_1, g_1) and (M_2, g_2) be semi-Riemannian manifolds and let $(\mathcal{F}, g_{2\mathcal{F}})$ be a semi-Riemannian foliation on (M_2, g_2). Then the equation

$$\|f_*{}^{\perp}\|^2 = \text{rank } \overline{f_*}^{\perp}$$

defined for the maps $f : (M_1, g_1) \to (M_2, g_2)$ by

$$\|f_*{}^{\perp}\|^2(p_1) = \text{rank } \overline{f_{*p_1}^{\perp}}$$

at each $p_1 \in M_1$ is called the *transversal generalized eikonal equation.* Hence weakly semi-Riemannian maps with respect to $(\mathcal{F}, g_{2\mathcal{F}})$ are the solutions of the transversal generalized eikonal equation.

Now we define semi-Riemannian maps with respect to a semi-Riemannian foliation which form a certain subset of the solutions of the transversal generalized eikonal equation with geometric nature.

Definition 5.1.4 Let $f : (M_1, g_1) \to (M_2, g_2)$ be a map and $(\mathcal{F}, g_{2\mathcal{F}})$ be a semi-Riemannian foliation on (M_2, g_2). Then f is called *semi-Riemannian with respect to* $(\mathcal{F}, g_{2\mathcal{F}})$ at $p_1 \in M_1$ if

$$\overline{f_{*p_1}^{\perp}} : (\overline{\mathcal{H}}^{\mathcal{F}}(p_1), \overline{g}_{1p_1 \overline{\mathcal{H}}^{\mathcal{F}}(p_1)}) \to (\overline{A_2}^{\mathcal{F}}(p_2), \overline{g}_{2p_2 \overline{A_2}^{\mathcal{F}}(p_2)})$$

is an (into) isometry, and f is called *semi-Riemannian with respect to* $(\mathcal{F}, g_{2\mathcal{F}})$ if f is semi-Riemannian with respect to $(\mathcal{F}, g_{2\mathcal{F}})$ at each $p_1 \in M_1$.

Remark 5.1.4 Note that, f is semi-Riemannian with respect to a semi-Riemannian foliation $(\mathcal{F}, g_{2\mathcal{F}})$ at $p_1 \in M_1$ is equivalent to $f_{*p_1}^{\perp}$ is semi-Euclidean. (See Definition 1.2.4). Also, if $(\mathcal{F}, g_{2\mathcal{F}})$ is the discrete foliation on (M_2, g_2) then the definition of a semi-Riemannian map with respect to $(\mathcal{F}, g_{2\mathcal{F}})$ reduces to the definition of a semi-Riemannian map. (See Remark 5.1.1).

Theorem 5.1.1 *If $f : (M_1, g_1) \to (M_2, g_2)$ is semi-Riemannian with respect to a semi-Riemannian foliation $(\mathcal{F}, g_{2\mathcal{F}})$ on (M_2, g_2) at $p_1 \in M_1$ then f is weakly semi-Riemannian with respect to $(\mathcal{F}, g_{2\mathcal{F}})$ at $p_1 \in M_1$. Hence, if f is semi-Riemannian with respect to $(\mathcal{F}, g_{2\mathcal{F}})$ then f is weakly semi-Riemannian with respect to $(\mathcal{F}, g_{2\mathcal{F}})$.*

Proof. Since $f_{*p_1}^{\perp}$ is semi-Euclidean by Remark 5.1.4, it follows from Theorem 1.2.2 that $\|f_{*p_1}^{\perp}\|^2 = \operatorname{rank} \overline{f_{*p_1}^{\perp}}$, that is, f is weakly semi-Riemannian with respect to $(\mathcal{F}, g_{2\mathcal{F}})$ at $p_1 \in M_1$. □

Remark 5.1.5 Let $f : (M_1, g_1) \to (M_2, g_2)$ be a semi-Riemannian map with respect to a semi-Riemannian foliation $(\mathcal{F}, g_{2\mathcal{F}})$ on (M_2, g_2) at $p_1 \in M_1$. Then it is worth emphasizing that $\overline{f_{*p_1}^{\perp}}$ is injective (yet may not be surjective) and hence, $\operatorname{rank} \overline{f_{*p_1}^{\perp}} = \operatorname{rank} f_{*p_1}^{\perp} - \dim L_1^{\mathcal{F}}(p_1)$. (See Remark 1.2.3). Also, if f is semi-Riemannian with respect to $(\mathcal{F}, g_{2\mathcal{F}})$ and M_1 is connected then, since f is weakly semi-Riemannian with respect to $(\mathcal{F}, g_{2\mathcal{F}})$, it follows from Remark 5.1.3 that the nondegenerate transversal rank of f with respect to $(\mathcal{F}, g_{2\mathcal{F}})$ is constant. But rank of $f_{*p_1}^{\perp}$ may change point to point.

Remark 5.1.6 Let (M_2, g_2) be a Riemannian manifold and $\mathcal{F} = \{S_\alpha\}_{\alpha \in A}$ be a foliation on (M_2, g_2). Then note that, since g_{2S_α} is a positive definite metric tensor on S_α for each $\alpha \in A$, $(\mathcal{F}, g_{2\mathcal{F}})$ is a semi-Riemannian foliation on (M_2, g_2), which we call a *Riemannian foliation on* (M_2, g_2). Now let (M_1, g_1) and (M_2, g_2) be Riemannian manifolds and $(\mathcal{F}, g_{2\mathcal{F}})$ be a Riemannian foliation on (M_2, g_2). We call a map $f : (M_1, g_1) \to (M_2, g_2)$ *Riemannian with respect to* $(\mathcal{F}, g_{2\mathcal{F}})$ *at* $p_1 \in M_1$ if f is a semi-Riemannian map with respect to $(\mathcal{F}, g_{2\mathcal{F}})$ at $p_1 \in M_1$. Then since we are dealing with positive definite inner product spaces, by making a completely analogous setting as in section 4.3 by replacing f_{*p_1} with $f_{*p_1}^{\perp}$, we obtain a characterization of Riemannian maps with respect to $(\mathcal{F}, g_{2\mathcal{F}})$ at $p_1 \in M_1$ as in Proposition 4.3.1. However, a Riemannian map with respect to a Riemannian foliation may not have locally constant rank although its transversal rank is locally constant. Hence in general, a Riemannian map with respect to a Riemannian foliation is not a subimmersion.

5.2 Semi-Riemannian Transversal Maps

Let $(\mathcal{F}, g_{2\mathcal{F}})$ be a semi-Riemannian foliation on (M_2, g_2) and $f : (M_1, g_1) \to (M_2, g_2)$ be a map. Also let $p_2 = f(p_1)$ for each $p_1 \in M_1$. We call f *transversal to* $(\mathcal{F}, g_{2\mathcal{F}})$ if

$$f_{*p_1}(T_{p_1} M_1) + T_{p_2}(M_2, \mathcal{F}) = T_{p_2} M_2$$

at each $p_1 \in M_1$. Then note that, the connected components of $f^{-1}(S_\alpha)$, $\alpha \in A$, are the leaves of a foliation on M_1 denoted by $f^*(\mathcal{F})$. (See [AMR, p. 337]). Also since

$$T_{p_1}(f^{-1}(S_\alpha)) = (f_{*p_1})^{-1}(T_{p_2}S_\alpha),$$

where S_α is the leaf of $\mathcal{F}$ passing through $p_2 \in M_2$ (cf. [AMR, p. 202]),

$$T_{p_1}(f^{-1}(S_\alpha)) = \ker f_{*p_1}^{\perp}.$$

It is worth emphasizing here that, if $f : (M_1, g_1) \to (M_2, g_2)$ is transversal to a semi-Riemannian foliation $(\mathcal{F}, g_{2_\mathcal{F}})$ on (M_2, g_2) then $f_{*p_1}^{\perp} : T_{p_1}M_1 \to T_{p_2}(M_2, \mathcal{F})^{\perp}$ is surjective at each $p_1 \in M_1$.

Also in particular, if we take $(\mathcal{F}, g_{2_\mathcal{F}})$ to be the discrete foliation on (M_2, g_2), that is, the leaves of $\mathcal{F}$ are the points of M_2, then a transversal map $f : (M_1, g_1) \to (M_2, g_2)$ to $(\mathcal{F}, g_{2_\mathcal{F}})$ is a submersion. Next we will generalize the notion of semi-Riemannian submersion to the notion of semi-Riemannian transversal map.

Definition 5.2.1 Let $f : (M_1, g_1) \to (M_2, g_2)$ be a map and $(\mathcal{F}, g_{2_\mathcal{F}})$ be a semi-Riemannian foliation on (M_2, g_2). If

(1) f is transversal to $(\mathcal{F}, g_{2_\mathcal{F}})$,

(2) f is semi-Riemannian with respect to $(\mathcal{F}, g_{2_\mathcal{F}})$,

then f is called a *semi-Riemannian transversal map to* $(\mathcal{F}, g_{2_\mathcal{F}})$.

In the next remark, we disclose the content of this definition. But in the first place, note that if $(\mathcal{F}, g_{2_\mathcal{F}})$ is the discrete foliation on (M_2, g_2) then Definition 5.2.1 reduces to Definition 4.2.1 of a semi-Riemannian submersion.

Remark 5.2.1 Let $(\mathcal{F}, g_{2_\mathcal{F}})$ be a semi-Riemannian foliation on (M_2, g_2) and $f : (M_1, g_1) \to (M_2, g_2)$ be a semi-Riemannian transversal map to $(\mathcal{F}, g_{2_\mathcal{F}})$, where M_1 is connected and $f(M_1)$ is an immersed submanifold of M_2. Then since f is a transversal map to $(\mathcal{F}, g_{2_\mathcal{F}})$,

$$\mathcal{V}^{\mathcal{F}} = \bigcup_{p_1 \in M_1} \mathcal{V}^{\mathcal{F}}(p_1) = \ker f_*{}^{\perp}$$

is a subbundle of TM_1 and hence,

$$\mathcal{H}^{\mathcal{F}} = \bigcup_{p_1 \in M_1} \mathcal{H}^{\mathcal{F}}(p_1) = (\ker f_*{}^{\perp})^{\perp}$$

is also a subbundle of TM_1, where $(\ker f_*{}^{\perp})^{\perp}$ is the orthogonal bundle to $\ker f_*{}^{\perp}$ defined as usual by $(\ker f_*{}^{\perp})^{\perp} = \bigcup_{p_1 \in M_1} (\ker f_{*p_1}^{\perp})^{\perp}$. Now observe that, since f is a semi-Riemannian transversal map to $(\mathcal{F}, g_{2\mathcal{F}})$ and M_1 is connected, it follows that $\dim L_1^{\mathcal{F}}(p_1) = \operatorname{rank} f_{*p_1}^{\perp} - \operatorname{rank} \overline{f_{*p_1}^{\perp}}$ is constant for all $p_1 \in M_1$. Thus

$$L_1^{\mathcal{F}} = \bigcup_{p_1 \in M_1} L_1^{\mathcal{F}}(p_1) = \mathcal{V}^{\mathcal{F}} \cap \mathcal{H}^{\mathcal{F}} = (\ker f_*{}^{\perp}) \cap (\ker f_*{}^{\perp})^{\perp}$$

is a subbundle of TM_1. Also note that, since $L_1^{\mathcal{F}}$ is a subbundle of TM_1,

$$\overline{\mathcal{H}}^{\mathcal{F}} = \bigcup_{p_1 \in M_1} \overline{\mathcal{H}}^{\mathcal{F}}(p_1)$$

is a vector bundle over M_1 and, by Theorem 2.1.2, the inner products $\overline{g_1}_{p_1 \overline{\mathcal{H}}^{\mathcal{F}}(p_1)}$ defined on $\overline{\mathcal{H}}^{\mathcal{F}}(p_1)$ at each $p_1 \in M_1$ define a metric tensor $\overline{g_1}_{\overline{\mathcal{H}}^{\mathcal{F}}}$ in $\overline{\mathcal{H}}^{\mathcal{F}}$ by

$$(\overline{g_1}_{\overline{\mathcal{H}}^{\mathcal{F}}})_{p_1} = \overline{g_1}_{p_1 \overline{\mathcal{H}}^{\mathcal{F}}(p_1)}.$$

Thus $(\overline{\mathcal{H}}^{\mathcal{F}}, \overline{g_1}_{\overline{\mathcal{H}}^{\mathcal{F}}})$ is a semi-Riemannian vector bundle over M_1 of some index. Hence it follows by a similar argument as in Remark 4.2.1 for semi-Riemannian submersions that the leaves of $f^*(\mathcal{F})$ are degenerate submanifolds of some constant type, provided that $\operatorname{rank} L_1^{\mathcal{F}} \neq 0$, and the leaves of $f^*(\mathcal{F})$ are semi-Riemannian submanifolds of some constant index, provided that $\operatorname{rank} L_1^{\mathcal{F}} = 0$.

On the other hand, since f is a transversal map to $(\mathcal{F}, g_{2\mathcal{F}})$, range $f_{*p_1}^{\perp} = T_{p_2}(M_2, \mathcal{F})^{\perp}$ at each $p_1 \in M_1$. Thus $L_2^{\mathcal{F}}(p_2) = \{0\}$ and therefore, $(A_2^{\mathcal{F}}(p_2), g_{2p_2 A_2^{\mathcal{F}}(p_2)})$ is canonically isometric to $(\overline{A_2}^{\mathcal{F}}(p_2), \overline{g_2}_{p_2 \overline{A_2}^{\mathcal{F}}(p_2)})$ by $\pi_{A_2^{\mathcal{F}}(p_2)}$ at each $p_2 \in f(M_1)$. Now by identifying $(A_2^{\mathcal{F}}(p_2), g_{2p_2 A_2^{\mathcal{F}}(p_2)})$ and $(\overline{A_2}^{\mathcal{F}}(p_2), \overline{g_2}_{p_2 \overline{A_2}^{\mathcal{F}}(p_2)})$ at each $p_2 \in f(M_1)$, we define

$$\overline{A_2}^{\mathcal{F}} = A_2^{\mathcal{F}} = \bigcup_{p_2 \in f(M_1)} A_2^{\mathcal{F}}(p_2) = T(M_2, \mathcal{F})^{\perp} |_{f(M_1)}.$$

Note that, since $f(M_1)$ is assumed to be an immersed submanifold of M_2, $\overline{A_2}^{\mathcal{F}} = A_2^{\mathcal{F}}$ is a vector bundle over $f(M_1)$.

Thus we can extend $\overline{f_{*p_1}^{\perp}}$ defined at each $p_1 \in M_1$ to a linear map

$$\overline{f_*^{\perp}} : \overline{\mathcal{H}}^{\mathcal{F}} \to \overline{A_2}^{\mathcal{F}} = A_2^{\mathcal{F}}$$

by

$$\overline{f_*^{\perp}}\,\overline{x} = \overline{f_{*p_1}^{\perp}}\,\overline{x} = f_{*p_1}^{\perp} x,$$

where $\overline{x} \in \overline{\mathcal{H}^{\mathcal{F}}}(p_1)$ and $x \in \mathcal{H}^{\mathcal{F}}(p_1)$ with $\pi_{\mathcal{H}^{\mathcal{F}}(p_1)}(x) = \overline{x}$. Then note that since $\overline{f_{*p_1}^{\perp}}$ is an (into) isometry at each $p_1 \in M_1$,

$$\overline{f_*^{\perp}} : (\overline{\mathcal{H}^{\mathcal{F}}}, \overline{g_1}_{\overline{\mathcal{H}}^{\mathcal{F}}}) \to (A_2^{\mathcal{F}}, g_{2T(M_2,\mathcal{F})^{\perp}|_{f(M_1)}})$$

is an (into) isometric linear map between semi-Riemannian vector bundles $(\overline{\mathcal{H}^{\mathcal{F}}}, \overline{g_1}_{\overline{\mathcal{H}}^{\mathcal{F}}})$ and $(A_2^{\mathcal{F}}, g_{2T(M_2,\mathcal{F})^{\perp}|_{f(M_1)}})$. Now let

$$\mathcal{B}^{\mathcal{F}} = \text{range } \overline{f_*^{\perp}}.$$

Observe that, if $\overline{f_{*p_1}^{\perp}}(\overline{\mathcal{H}^{\mathcal{F}}}(p_1)) = \overline{f_{*q_1}^{\perp}}(\overline{\mathcal{H}^{\mathcal{F}}}(q_1))$ for all p_1, $q_1 \in M_1$ with $f(p_1) = f(q_1)$ then $\mathcal{B}^{\mathcal{F}} = \text{range } \overline{f_*^{\perp}}$ is a subbundle of $A_2^{\mathcal{F}}$. In this case, we call $\mathcal{B}^{\mathcal{F}}$ an $\overline{f_*^{\perp}}$-*subbundle of* $A_2^{\mathcal{F}}$. Note that, if $\mathcal{B}^{\mathcal{F}}$ is a $\overline{f_*^{\perp}}$-*subbundle of* $A_2^{\mathcal{F}}$ then $(\mathcal{B}^{\mathcal{F}}, g_{2\mathcal{B}^{\mathcal{F}}})$ is a semi-Riemannian subbundle of $(A_2^{\mathcal{F}}, g_{2T(M_2,\mathcal{F})^{\perp}|_{f(M_1)}})$ and

$$\overline{f_*^{\perp}} : (\overline{\mathcal{H}^{\mathcal{F}}}, \overline{g_1}_{\overline{\mathcal{H}}^{\mathcal{F}}}) \to (\mathcal{B}^{\mathcal{F}}, g_{2\mathcal{B}^{\mathcal{F}}})$$

is an isometric bundle map, where $g_{2\mathcal{B}^{\mathcal{F}}}$ is the restriction of $g_{2T(M_2,\mathcal{F})^{\perp}|_{f(M_1)}}$ to $\mathcal{B}^{\mathcal{F}}$.

Definition 5.2.2 Let $f : (M_1, g_1) \to (M_2, g_2)$ be a map and $(\mathcal{F}, g_{2\mathcal{F}})$ be a semi-Riemannian foliation on (M_2, g_2). Then f is called *normal* to $(\mathcal{F}, g_{2\mathcal{F}})$ if $f_{*p_1} = f_{*p_1}^{\perp}$ at each $p_1 \in M_1$, that is, $f_{*p_1}(T_{p_1}M_1) \perp T_{p_2}(M_2, \mathcal{F})$ at each $p_1 \in M_1$.

Remark 5.2.2 Note that a semi-Riemannian submersion $f : (M_1, g_1) \to (M_2, g_2)$ is normal semi-Riemannian transversal map to the discrete foliation on (M_2, g_2).

Remark 5.2.3 Let $(\mathcal{F}, g_{2\mathcal{F}})$ be a semi-Riemannian foliation on (M_2, g_2) and let $f : (M_1, g_1) \to (M_2, g_2)$ be a normal semi-Riemannian transversal map to $(\mathcal{F}, g_{2\mathcal{F}})$, where M_1 is connected, $f(M_1)$ is an immersed submanifold of M_2 and $\mathcal{B}^{\mathcal{F}}$ is an $\overline{f_*^{\perp}}$-subbundle of $A_2^{\mathcal{F}}$. Then note that, since $f_{*p_1} = f_{*p_1}^{\perp}$ at each $p_1 \in M_1$, range $f_{*p_1} = T_{p_2}(M, \mathcal{F})^{\perp}$ at each $p_1 \in M_1$, and hence f is of constant rank on M_1. Thus f is a subimmersion and it follows that range $f_{*p_1} = T_{p_2}(f(M_1))$ at each $p_1 \in M_1$. (See [AMR, p. 206], [F1, Theorem 4.3]). Therefore $T(f(M_1)) = T(M, \mathcal{F})^{\perp} |_{f(M_1)} = A_2^{\mathcal{F}}$ and, if $g_{2f(M_1)}$ is the restriction of g_1 to $f(M_1)$, then $(f(M_1), g_{2f(M_1)})$ is an immersed semi-Riemannian submanifold of (M_2, g_2). Thus $(\mathcal{B}^{\mathcal{F}}, g_{2\mathcal{B}^{\mathcal{F}}})$ is a semi-Riemannian subbundle of the (semi-Riemannian) tangent bundle of $f(M_1)$. In the next proposition, by integrability of $\mathcal{B}^{\mathcal{F}}$, we mean the integrability of $\mathcal{B}^{\mathcal{F}}$ as a subbundle of $T(f(M_1))$. Also note that $f(M_1)$ is transversal to the leaves of $\mathcal{F}$ and, since $f(M_1)$ and the leaves of $\mathcal{F}$ have complementary dimensions, intersection of $f(M_1)$ with each leaf of $\mathcal{F}$ is either empty or a single point. Hence, it can be easily observed now that $f : (M_1, g_1) \to (f(M_1), g_{2f(M_1)})$ is a semi-Riemannian submersion with $f^*(\mathcal{F})$ as the connected components of its fibers.

Proposition 5.2.1 *Let $(\mathcal{F}, g_{2\mathcal{F}})$ be a semi-Riemannian foliation on (M_2, g_2) and let $f : (M_1, g_1) \to (M_2, g_2)$ be a normal semi-Riemannian transversal map to $(\mathcal{F}, g_{2\mathcal{F}})$, where M_1 is connected, $f(M_1)$ is an immersed submanifold of M_2 and $\mathcal{B}^{\mathcal{F}}$ is an $\overline{f_*}^{\perp}$-subbundle of $A_2^{\mathcal{F}}$. If $A_1^{\mathcal{F}} = (\ker f_*{}^{\perp}) + (\ker f_*{}^{\perp})^{\perp}$ is integrable then $\mathcal{B}^{\mathcal{F}} = \operatorname{range} \overline{f_*}^{\perp}$ is integrable and the integral manifolds of $\mathcal{B}^{\mathcal{F}}$ are semi-Riemannian submanifolds of (M_2, g_2) of the same index.*

Proof. First note that $f_*(A_1^{\mathcal{F}}) = \mathcal{B}^{\mathcal{F}}$. Now let $X, Y \in \Gamma\mathcal{B}^{\mathcal{F}}$ and let $\tilde{X}, \tilde{Y} \in \Gamma A_1^{\mathcal{F}}$ be lifts of X, Y to $A_1^{\mathcal{F}}$, respectively, that is, $f_*\tilde{X} = X \circ f$ and $f_*\tilde{Y} = Y \circ f$. Then since $f_*([\tilde{X}, \tilde{Y}]) = [X, Y] \circ f$ and $[\tilde{X}, \tilde{Y}] \in \Gamma A_1^{\mathcal{F}}$, it follows that $[X, Y] \in \Gamma\mathcal{B}^{\mathcal{F}}$. That is $\mathcal{B}^{\mathcal{F}}$ is integrable. Thus by Remark 5.2.1, integral manifolds of $\mathcal{B}^{\mathcal{F}}$ are semi-Riemannian submanifolds of (M_2, g_2) of the same index. □

Definition 5.2.3 Let $(\mathcal{F}, g_{2\mathcal{F}})$ be a semi-Riemannian foliation on (M_2, g_2) and let $f : (M_1, g_1) \to (M_2, g_2)$ be a semi-Riemannian transversal map to $(\mathcal{F}, g_{2\mathcal{F}})$. Then f is called *regular* if $A_1^{\mathcal{F}} = (\ker f_*{}^{\perp}) + (\ker f_*{}^{\perp})^{\perp}$ is a totally geodesic distribution, that is, $\overset{1}{\nabla}_X Y \in \Gamma A_1^{\mathcal{F}}$ for every $X, Y \in \Gamma A_1^{\mathcal{F}}$.

Remark 5.2.4 As in Remark 4.2.2, if $f : (M_1, g_1) \to (M_2, g_2)$ is a regular, normal semi-Riemannian transversal map to a semi-Riemannian foliation $(\mathcal{F}, g_{2\mathcal{F}})$ on (M_2, g_2), where M_1 is connected, $f(M_1)$ is an immersed submanifold of M_2 and $\mathcal{B}^{\mathcal{F}}$ is an $\overline{f_*}^{\perp}$-subbundle of $A_2^{\mathcal{F}}$, then the regularity of f implies that $A_1^{\mathcal{F}}$ is integrable, and hence $\mathcal{B}^{\mathcal{F}}$ is integrable by Proposition 5.2.1. Also if the leaves of $f^*(\mathcal{F})$ are degenerate submanifolds then they are stationary and irrotational. (See Proposition A.2.5).

Here it is worth giving an example of a regular, normal semi-Riemannian transversal map to a semi-Riemannian foliation to have a picture of it in mind. Let $(M_1, g_1) = (\mathbf{R}, dt \otimes dt)$ and $(M_2, g_2) = (\mathbf{R}^2, -dx^1 \otimes dx^1 + dx^2 \otimes dx^2)$. Also let $\mathcal{F} = \{S_\alpha = \{x^2 = \alpha / \alpha \in \mathbf{R}\}\}_{\alpha \in \mathbf{R}}$ be a foliation on (M_2, g_2). Note that the induced metric tensor g_{2S_α} on each S_α is negative definite and hence $(\mathcal{F}, g_{2\mathcal{F}})$ is a semi-Riemannian foliation on (M_2, g_2). Now let $f : (M_1, g_1) \to (M_2, g_2)$ be defined by $f(t) = (0, t)$. Then it is easy to see that f is a regular, normal semi-Riemannian transversal map to $(\mathcal{F}, g_{2\mathcal{F}})$, where M_1 is connected, $f(M_1)$ is an (immersed) submanifold of M_2 and $\mathcal{B}^{\mathcal{F}}$ is an $\overline{f_*}^{\perp}$-subbundle of $A_2^{\mathcal{F}}$. Notice also that $f^*(\mathcal{F})$ is the discrete foliation on (M_1, g_1).

The semi-Riemannian nature of a semi-Riemannian transversal map enables us to compare curvatures of manifolds. Next we will state a fundamental equation of a semi-Riemannian transversal map as such a comparison of curvatures of the manifolds.

Let $(\mathcal{F}, g_{2\mathcal{F}})$ be a semi-Riemannian foliation on (M_2, g_2) and $f : (M_1, g_1) \to (M_2, g_2)$ be a regular, normal semi-Riemannian transversal map to $(\mathcal{F}, g_{2\mathcal{F}})$, where

M_1 is connected, $f(M_1)$ is an immersed submanifold of M_2 and $\mathcal{B}^{\mathcal{F}}$ is an $\overline{f_*}^{\perp}$-subbundle of $A_2^{\mathcal{F}}$. Let $\overset{2}{D}$ denote the collection of the Levi-Civita connections of the (semi-Riemannian) integral manifolds (N_2, g_{2N_2}) of $\mathcal{B}^{\mathcal{F}}$, where g_{2N_2} is the restriction of g_2 to integral manifold N_2 of $\mathcal{B}^{\mathcal{F}}$. Also let $\overset{2}{R}_{N_2}$ be the collection of the curvature tensors of integral manifolds (N_2, g_{2N_2}) of $\mathcal{B}^{\mathcal{F}}$. Let $(\mathcal{H}^{\mathcal{F}})'$ be a complementary subbundle to $L_1^{\mathcal{F}}$ in $\mathcal{H}^{\mathcal{F}}$ and $(\mathcal{V}^{\mathcal{F}})'$ be a complementary subbundle to $L_1^{\mathcal{F}}$ in $\mathcal{V}^{\mathcal{F}}$. Then note that

$$A_1^{\mathcal{F}} = (\mathcal{H}^{\mathcal{F}})' \oplus L_1^{\mathcal{F}} \oplus (\mathcal{V}^{\mathcal{F}})'$$

is a direct sum decomposition of $A_1^{\mathcal{F}}$. Thus, if $U \in \Gamma A_1^{\mathcal{F}}$ then U can be written uniquely as

$$U = U^{\perp} + U^0 + U^T,$$

where $U^{\perp}$, U^0 and U^T are the components of U in $\Gamma(\mathcal{H}^{\mathcal{F}})'$, $\Gamma L_1^{\mathcal{F}}$ and $\Gamma(\mathcal{V}^{\mathcal{F}})'$, respectively. Also note that if $g_{1(\mathcal{H}^{\mathcal{F}})'}$ is the restriction of g_1 to $(\mathcal{H}^{\mathcal{F}})'$, then $((\mathcal{H}^{\mathcal{F}})', g_{1(\mathcal{H}^{\mathcal{F}})'})$ is a semi-Riemannian subbundle of (TM_1, g_1) and

$$f_*^{\perp} = f_* : ((\mathcal{H}^{\mathcal{F}})', g_{1(\mathcal{H}^{\mathcal{F}})'}) \to (\mathcal{B}^{\mathcal{F}}, g_{2\mathcal{B}^{\mathcal{F}}})$$

becomes an isometric bundle map. Thus in particular, for every $X \in \Gamma\mathcal{B}^{\mathcal{F}}$, there is a unique lift $\tilde{X}$ of X to $(\mathcal{H}^{\mathcal{F}})'$, that is, $\tilde{X} \in \Gamma(\mathcal{H}^{\mathcal{F}})'$ with $f_*^{\perp}\tilde{X} = f_*\tilde{X} = X \circ f$.

Lemma 5.2.1 *Let $(\mathcal{F}, g_{2\mathcal{F}})$ be a semi-Riemannian foliation on (M_2, g_2) and $f : (M_1, g_1) \to (M_2, g_2)$ be a regular, normal semi-Riemannian transversal map to $(\mathcal{F}, g_{2\mathcal{F}})$, where M_1 is connected, $f(M_1)$ is an immersed submanifold of M_2 and $\mathcal{B}^{\mathcal{F}}$ is an $\overline{f_*}^{\perp}$-subbundle of $A_2^{\mathcal{F}}$. Also let $\tilde{X}$, $\tilde{Y} \in \Gamma(\mathcal{H}^{\mathcal{F}})'$ be the lifts of X, $Y \in \Gamma\mathcal{B}^{\mathcal{F}}$, respectively. Then*

(1) $g_1(\tilde{X}, \tilde{Y}) = g_2(X, Y) \circ f$.

(2) $[\tilde{X}, \tilde{Y}]^{\perp} = [X, Y]^{\sim}$.

(3) $(\overset{1}{\nabla}_{\tilde{X}} \tilde{Y})^{\perp} = (\overset{2}{D}_X Y)^{\sim}$.

(4) $(\overset{1}{\nabla}_{\tilde{X}} \tilde{Y})^T = \frac{1}{2}[\tilde{X}, \tilde{Y}]^T$.

Proof. (1) Since $f_* : ((\mathcal{H}^{\mathcal{F}})', g_{1(\mathcal{H}^{\mathcal{F}})'}) \to (\mathcal{B}^{\mathcal{F}}, g_{2\mathcal{B}^{\mathcal{F}}})$ is a bundle isometry,

$$g_1(\tilde{X}, \tilde{Y}) = g_2(f_*\tilde{X}, f_*\tilde{Y}) = g_2(X \circ f, Y \circ f) = g_2(X, Y) \circ f.$$

(2) Since $A_1^{\mathcal{F}}$ is integrable and $f_*[\tilde{X},\tilde{Y}] = [X,Y]\circ f$,

$$[X,Y]\circ f = f_*[\tilde{X},\tilde{Y}] = f_*([\tilde{X},\tilde{Y}]^{\perp} + [\tilde{X},\tilde{Y}]^0 + [\tilde{X},\tilde{Y}]^T) = f_*([\tilde{X},\tilde{Y}]^{\perp}).$$

Hence $[\tilde{X},\tilde{Y}]^{\perp} = [X,Y]^{\sim}$.

(3) It suffices to show that for every $\tilde{Z} \in \Gamma(\mathcal{H}^{\mathcal{F}})'$ which is lift of $Z \in \Gamma\mathcal{B}^{\mathcal{F}}$, $g_1(\overset{1}{\nabla}_{\tilde{X}}\tilde{Y},\tilde{Z}) = g_1((\overset{2}{D}_X Y)^{\sim},\tilde{Z})$. Then since $A_1^{\mathcal{F}}$ is totally geodesic and $g_{1(\mathcal{H}^{\mathcal{F}})'}$ is nondegenerate, the claim follows.

Let $\tilde{Z} \in \Gamma(\mathcal{H}^{\mathcal{F}})'$ be the lift of $Z \in \Gamma\mathcal{B}^{\mathcal{F}}$, that is, $f_*\tilde{Z} = Z\circ f$. Then by using (1), (2) and the Koszul formula (see Theorem 2.2.2),

$$\begin{aligned} g_1(\overset{1}{\nabla}_{\tilde{X}}\tilde{Y},\tilde{Z}) &= \tfrac{1}{2}\Big[\tilde{X}g_1(\tilde{Y},\tilde{Z}) + \tilde{Y}g_1(\tilde{Z},\tilde{X}) - \tilde{Z}g_1(\tilde{X},\tilde{Y}) \\ &\quad + g_1(\tilde{Z},[\tilde{X},\tilde{Y}]) + g_1(\tilde{Y},[\tilde{Z},\tilde{X}]) - g_1(\tilde{X},[\tilde{Y},\tilde{Z}])\Big] \\ &= \tfrac{1}{2}\Big[\tilde{X}g_1(\tilde{Y},\tilde{Z}) + \tilde{Y}g_1(\tilde{Z},\tilde{X}) - \tilde{Z}g_1(\tilde{X},\tilde{Y}) \\ &\quad + g_1(\tilde{Z},[\tilde{X},\tilde{Y}]^{\perp}) + g_1(\tilde{Y},[\tilde{Z},\tilde{X}]^{\perp}) - g_1(\tilde{X},[\tilde{Y},\tilde{Z}]^{\perp})\Big] \\ &= \tfrac{1}{2}[Xg_2(Y,Z) + Yg_2(Z,X) - Zg_2(X,Y) \\ &\quad + g_2(Z,[X,Y]) + g_2(Y,[Z,X]) - g_2(X,[Y,Z])]\circ f \\ &= g_2(\overset{2}{D}_X Y, Z)\circ f \\ &= g_1((\overset{2}{D}_X Y)^{\sim},\tilde{Z}). \end{aligned}$$

(4) Let $U \in \Gamma(\mathcal{V}^{\mathcal{F}})'$. Then from the Koszul formula, we have

$$\begin{aligned} g_1(\overset{1}{\nabla}_{\tilde{X}}\tilde{Y},U) &= \tfrac{1}{2}\Big[\tilde{X}g_1(\tilde{Y},U) + \tilde{Y}g_1(U,\tilde{X}) - Ug_1(\tilde{X},\tilde{Y}) \\ &\quad + g_1(U,[\tilde{X},\tilde{Y}]) + g_1(\tilde{Y},[U,\tilde{X}]) - g_1(\tilde{X},[\tilde{Y},U])\Big]. \end{aligned}$$

But since $g_1(\tilde{X},\tilde{Y}) = g_2(X,Y)\circ f$ and $f_*U = 0\circ f$, $Ug_1(\tilde{X},\tilde{Y}) = 0$, and since $U \perp \tilde{X}$ and $U \perp \tilde{Y}$,

$$\tilde{X}g_1(\tilde{Y},U) = 0 = \tilde{Y}g_1(U,\tilde{X}).$$

Also, since $A_1^{\mathcal{F}}$ is integrable and $f_*U = 0 \circ f$,

$$\begin{aligned} g_1(\tilde{Y}, [U, \tilde{X}]) &= g_1(\tilde{Y}, [U, \tilde{X}]^{\perp} + [U, \tilde{X}]^0 + [U, \tilde{X}]^T) \\ &= g_1(\tilde{Y}, [U, \tilde{X}]^{\perp}) \\ &= g_2(Y \circ f, f_*[U, \tilde{X}]^{\perp}) \\ &= g_2(Y \circ f, f_*([U, \tilde{X}]^{\perp} + [U, \tilde{X}]^0 + [U, \tilde{X}]^T)) \\ &= g_2(Y \circ f, f_*[U, \tilde{X}]) \\ &= g_2(Y, [0, X]) \circ f = 0 \end{aligned}$$

and similarly

$$g_1(\tilde{X}, [\tilde{Y}, U]) = 0.$$

Thus we have

$$g_1(\overset{1}{\nabla}_{\tilde{X}} \tilde{Y}, U) = \frac{1}{2} g_1(U, [\tilde{X}, \tilde{Y}]) = \frac{1}{2} g_1([\tilde{X}, \tilde{Y}], U)$$

for every $U \in \Gamma(\mathcal{V}^{\mathcal{F}})'$. Hence since $A_1^{\mathcal{F}}$ is a totally geodesic distribution and the restriction of g_1 to $(\mathcal{V}^{\mathcal{F}})'$ is nondegenerate, the claim follows. □

Theorem 5.2.1 [K4] *Let $(\mathcal{F}, g_{2\mathcal{F}})$ be a semi-Riemannian foliation on (M_2, g_2) and $f : (M_1, g_1) \to (M_2, g_2)$ be a regular, normal semi-Riemannian transversal map to $(\mathcal{F}, g_{2\mathcal{F}})$, where M_1 is connected, $f(M_1)$ is an immersed submanifold of M_2 and $\mathcal{B}^{\mathcal{F}}$ is an $\overline{f_*}^{\perp}$-subbundle of $A_2^{\mathcal{F}}$. If $\tilde{X}, \tilde{Y} \in \Gamma(\mathcal{H}^{\mathcal{F}})'$ are the lifts of $X, Y \in \Gamma\mathcal{B}^{\mathcal{F}}$ spanning nondegenerate planes at each point then*

$$\begin{aligned} K_{N_2}(\mathrm{span}\{X, Y\}) \circ f &= K_{M_1}(\mathrm{span}\{\tilde{X}, \tilde{Y}\}) \\ &\quad + \tfrac{3}{4} g_1([\tilde{X}, \tilde{Y}]^T, [\tilde{X}, \tilde{Y}]^T) / Q(\tilde{X}, \tilde{Y}), \end{aligned}$$

where K_{N_2} is the sectional curvature of the integral manifolds (N_2, g_{2N_2}) of $\mathcal{B}^{\mathcal{F}}$, K_{M_1} is the sectional curvature of (M_1, g_1) and $Q(\tilde{X}, \tilde{Y}) = g_1(\tilde{X}, \tilde{X}) g_1(\tilde{Y}, \tilde{Y}) - g_1(\tilde{X}, \tilde{Y})^2$. (See Definition 2.3.3).

Proof. First note that, by Lemma 5.2.1,

$$\begin{aligned} \overset{1}{\nabla}_{\tilde{X}} \tilde{Y} &= (\overset{1}{\nabla}_{\tilde{X}} \tilde{Y})^{\perp} + (\overset{1}{\nabla}_{\tilde{X}} \tilde{Y})^0 + (\overset{1}{\nabla}_{\tilde{X}} \tilde{Y})^T \\ &= (\overset{2}{D}_X Y)^{\tilde{}} + (\overset{1}{\nabla}_{\tilde{X}} \tilde{Y})^0 + \tfrac{1}{2} [\tilde{X}, \tilde{Y}]^T. \end{aligned}$$

Again by Lemma 5.2.1–(4) (also see its proof), since $f_* U = 0 \circ f$ for $U \in \Gamma(\mathcal{V}^{\mathcal{F}})'$,

$$\begin{aligned}
g_1(\overset{1}{\nabla}_U \tilde{X}, \tilde{Y}) &= g_1(\overset{1}{\nabla}_{\tilde{X}} U, \tilde{Y}) + g_1([U, \tilde{X}], \tilde{Y}) \\
&= g_1(\overset{1}{\nabla}_{\tilde{X}} U, \tilde{Y}) \\
&= \tilde{X} g_1(U, \tilde{Y}) - g_1(U, \overset{1}{\nabla}_{\tilde{X}} \tilde{Y}) \\
&= -g_1(U, \overset{1}{\nabla}_{\tilde{X}} \tilde{Y}) \\
&= -g_1(U, (\overset{1}{\nabla}_{\tilde{X}} \tilde{Y})^T) \\
&= -\tfrac{1}{2} g_1([\tilde{X}, \tilde{Y}]^T, U)
\end{aligned} \tag{5.1}$$

for $U \in \Gamma(\mathcal{V}^{\mathcal{F}})'$. Also note that, in particular, if $U' \in \Gamma L_1^{\mathcal{F}}$ then $g_1(\overset{1}{\nabla}_{U'} \tilde{X}, \tilde{Y}) = 0$.

Now let $\tilde{Z}$, $\tilde{W} \in \Gamma(\mathcal{H}^{\mathcal{F}})'$ be the lifts of Z, $W \in \Gamma \mathcal{B}^{\mathcal{F}}$, respectively. Then by Lemma 5.2.1, since

$$\begin{aligned}
\tilde{X} g_1(\overset{1}{\nabla}_{\tilde{Y}} \tilde{Z}, \tilde{W}) &= \tilde{X} g_1((\overset{1}{\nabla}_{\tilde{Y}} \tilde{Z})^{\perp}, \tilde{W}) \\
&= \tilde{X} g_1((\overset{2}{D}_Y Z)^{\sim}, \tilde{W}) \\
&= (X g_2(\overset{2}{D}_Y Z, W)) \circ f,
\end{aligned}$$

we have

$$\begin{aligned}
g_1(\overset{1}{\nabla}_{\tilde{X}} \overset{1}{\nabla}_{\tilde{Y}} \tilde{Z}, \tilde{W}) &= \tilde{X} g_1(\overset{1}{\nabla}_{\tilde{Y}} \tilde{Z}, \tilde{W}) - g_1(\overset{1}{\nabla}_{\tilde{Y}} \tilde{Z}, \overset{1}{\nabla}_{\tilde{X}} \tilde{W}) \\
&= (X g_2(\overset{2}{D}_Y Z, W)) \circ f \\
&\quad - g_1((\overset{2}{D}_Y Z)^{\sim} + (\overset{1}{\nabla}_{\tilde{Y}} \tilde{Z})^0 + \tfrac{1}{2}[\tilde{Y}, \tilde{Z}]^T, \\
&\quad (\overset{2}{D}_X W)^{\sim} + (\overset{1}{\nabla}_{\tilde{X}} \tilde{W})^0 + \tfrac{1}{2}[\tilde{X}, \tilde{W}]^T)
\end{aligned}$$

and thus,

$$
\begin{aligned}
g_1(\overset{1}{\nabla}_{\tilde{X}}\overset{1}{\nabla}_{\tilde{Y}}\tilde{Z},\tilde{W}) &= (Xg_2(\overset{2}{D}_Y Z,W))\circ f - g_1((\overset{2}{D}_Y Z)\tilde{}, (\overset{2}{D}_X W)\tilde{}) \\
&\quad -\tfrac{1}{4}g_1([\tilde{Y},\tilde{Z}]^T,[\tilde{X},\tilde{W}]^T) \\
&= (Xg_2(\overset{2}{D}_Y Z,W))\circ f - g_2(\overset{2}{D}_Y Z,\overset{2}{D}_X W)\circ f \\
&\quad -\tfrac{1}{4}g_1([\tilde{Y},\tilde{Z}]^T,[\tilde{X},\tilde{W}]^T) \\
&= g_2(\overset{2}{D}_X\overset{2}{D}_Y Z,W)\circ f - \tfrac{1}{4}g_1([\tilde{Y},\tilde{Z}]^T,[\tilde{X},\tilde{W}]^T).
\end{aligned}
$$

Similarly,

$$
g_1(\overset{1}{\nabla}_{\tilde{Y}}\overset{1}{\nabla}_{\tilde{X}}\tilde{Z},\tilde{W}) = g_2(\overset{2}{D}_Y\overset{2}{D}_X Z,W)\circ f - \frac{1}{4}g_1([\tilde{X},\tilde{Z}]^T,[\tilde{Y},\tilde{W}]^T).
$$

Also by Lemma 5.2.1, (5.1) and $g_1(\overset{1}{\nabla}_{U'}\tilde{Z},\tilde{W})=0$ for $U'\in\Gamma L_1^{\mathcal{F}}$,

$$
\begin{aligned}
g_1(\overset{1}{\nabla}_{[\tilde{X},\tilde{Y}]}\tilde{Z},\tilde{W}) &= g_1(\overset{1}{\nabla}_{[\tilde{X},\tilde{Y}]^\perp}\tilde{Z},\tilde{W}) + g_1(\nabla_{[\tilde{X},\tilde{Y}]^o}\tilde{Z},\tilde{W}) \\
&\quad +g_1(\overset{1}{\nabla}_{[\tilde{X},\tilde{Y}]^T}\tilde{Z},\tilde{W}) \\
&= g_1(\overset{1}{\nabla}_{[\tilde{X},\tilde{Y}]^\perp}\tilde{Z},\tilde{W}) + g_1(\overset{1}{\nabla}_{[\tilde{X},\tilde{Y}]^T}\tilde{Z},\tilde{W}) \\
&= g_1((\overset{1}{\nabla}_{[\tilde{X},\tilde{Y}]^\perp}\tilde{Z})^\perp,\tilde{W}) + g_1(\overset{1}{\nabla}_{[\tilde{X},\tilde{Y}]^T}\tilde{Z},\tilde{W}) \\
&= g_2(\overset{2}{D}_{[X,Y]}Z,W)\circ f - \tfrac{1}{2}g_1([\tilde{Z},\tilde{W}]^T,[\tilde{X},\tilde{Y}]^T).
\end{aligned}
$$

Then, it follows that

$$
\begin{aligned}
g_1(\overset{1}{R}(\tilde{X},\tilde{Y})\tilde{Z},\tilde{W}) &= g_1(\overset{1}{\nabla}_{\tilde{X}}\overset{1}{\nabla}_{\tilde{Y}}\tilde{Z},\tilde{W}) - g_1(\overset{1}{\nabla}_{\tilde{Y}}\overset{1}{\nabla}_{\tilde{X}}\tilde{Z},\tilde{W}) \\
&\quad -g_1(\overset{1}{\nabla}_{[\tilde{X},\tilde{Y}]}\tilde{Z},\tilde{W}) \\
&= g_2(\overset{2}{D}_X\overset{2}{D}_Y Z,W)\circ f - \tfrac{1}{4}g_1([\tilde{Y},\tilde{Z}]^T,[\tilde{X},\tilde{W}]^T) \\
&\quad -g_2(\overset{2}{D}_Y\overset{2}{D}_X Z,W)\circ f + \tfrac{1}{4}g_1([\tilde{X},\tilde{Z}]^T,[\tilde{Y},\tilde{W}]^T) \\
&\quad -g_2(\overset{2}{D}_{[X,Y]}Z,W)\circ f + \tfrac{1}{2}g_1([\tilde{Z},\tilde{W}]^T,[\tilde{X},\tilde{Y}]^T)
\end{aligned}
$$

and thus,

$$\begin{aligned} g_1(\overset{1}{R}(\tilde{X},\tilde{Y})\tilde{Z},\tilde{W}) &= g_2(\overset{2}{R_{N_2}}(X,Y)Z,W) - \tfrac{1}{4}g_1([\tilde{Y},\tilde{Z}]^T,[\tilde{X},\tilde{W}]^T) \\ &\quad + \tfrac{1}{4}g_1([\tilde{X},\tilde{Z}]^T,[\tilde{Y},\tilde{W}]^T) + \tfrac{1}{2}g_1([\tilde{Z},\tilde{W}]^T,[\tilde{X},\tilde{Y}]^T). \end{aligned}$$

Hence by setting $\tilde{Z} = \tilde{Y}$ and $\tilde{W} = \tilde{X}$, we obtain

$$\begin{aligned} g_1(\overset{1}{R}(\tilde{X},\tilde{Y})\tilde{Z},\tilde{W}) &= g_2(\overset{2}{R_{N_2}}(X,Y)Z,W) \circ f \\ &\quad - \tfrac{3}{4}g_1([\tilde{X},\tilde{Y}]^T,[\tilde{X},\tilde{Y}]^T). \end{aligned}$$

Thus since $Q(\tilde{X},\tilde{Y}) = Q(X,Y) \circ f$, by dividing the above equation to $Q(\tilde{X},\tilde{Y})$, the claim follows. □

Remark 5.2.5 Let $f : (M_1, g_1) \to (M_2, g_2)$ be a regular semi-Riemannian submersion, where M_1 is connected and $\mathcal{B}$ is an $\overline{f_*}$-subbundle of A_2. First note that, since f is a submersion, $f(M_1)$ is an (open) submanifold of M_2. Thus f is a regular, normal semi-Riemannian transversal map to the discrete foliation $\mathcal{F}$ on (M_2, g_2), where M_1 is connected, $f(M_1)$ is an immersed submanifold of M_2 and $\mathcal{B}^{\mathcal{F}}$ is an $\overline{f_*}^{\perp}$-subbundle of $A_2^{\mathcal{F}}$, by observing that $\overline{f_{*p_1}}^{\perp} = \overline{f_{*p_1}}$ at each $p_1 \in M_1$. Thus in this case, Theorem 4.2.1 becomes a special case of Theorem 5.2.1.

Chapter 6

Semi-Riemannian Eikonal Equations and The Semi-Riemannian Regular Interval Theorem

In this chapter we introduce semi-Riemannian eikonal equations and investigate the consequences of the existence of their solutions.

To understand the origin of semi-Riemannian eikonal equations, let us first compute the square norm of a map $f : (M, g) \to (\mathbf{R}, kdt \otimes dt)$, where k is either of -1 or 1. For this, let $\{X_1, \ldots, X_n\}$ be an orthonormal local frame, where $\dim M = n$. Then

$$\begin{aligned} \|f_*\|^2 &= \sum_{i=1}^{n} g(X_i, X_i)(kdt \otimes dt)(f_*X_i, f_*X_i) \\ &= \sum_{i=1}^{n} g(X_i, X_i)(kdt \otimes dt)(g(\nabla f, X_i)\frac{d}{dt} \circ f, g(\nabla f, X_i)\frac{d}{dt} \circ f) \\ &= \sum_{i=1}^{n} g(X_i, X_i) g(\nabla f, X_i)^2 (kdt \otimes dt)(\frac{d}{dt} \circ f, \frac{d}{dt} \circ f) \\ &= kg(\nabla f, \nabla f). \end{aligned}$$

Hence the generalized eikonal equation for $f : (M, g) \to (\mathbf{R}, kdt \otimes dt)$ becomes

$$g(\nabla f, \nabla f) = k \operatorname{rank} \overline{f_*}.$$

But since rank $\overline{f_*} \leq 1$, actually a weakly semi-Riemannian map $f : (M, g) \to (\mathbf{R}, kdt \otimes dt)$ satisfies the equation

$$g(\nabla f, \nabla f) = k_0,$$

where k_0 is either of -1, 0 or 1. We call such equations

$$g(\nabla f, \nabla f) = k_0, \quad k_0 = -1, 0, 1$$

defined for the maps $f : (M, g) \to \mathbf{R}$ *semi-Riemannian eikonal equations of* (M, g). Clearly, solutions of semi-Riemannian eikonal equations of (M, g) are weakly semi-Riemannian maps $f : (M, g) \to (\mathbf{R}, kdt \otimes dt)$, where k is either of -1 or 1.

In section 6.1, we investigate the relations among the solutions of semi-Riemannian eikonal equations, semi-Riemannian submersions, and semi-Riemannian maps. Also, we emphasize the fact that semi-Riemannian eikonal equations do have local solutions but may not have global solutions. In section 6.2, we investigate the relations between solutions of semi-Riemannian eikonal equations and affine maps. We show that an affine map differs from a solution of a semi-Riemannian eikonal equation only by a multiplicative constant. Also we obtain some sufficient conditions under which a solution of a semi-Riemannian equation is an affine map. Finally in section 6.3, we state the semi-Riemannian regular interval theorem. Later in Chapter 7, we will make applications of this theorem to some splitting theorems in semi-Riemannian, Lorentzian, and Riemannian geometry.

Throughout this chapter, let (M, g) be a semi-Riemannian manifold of index ν and dimension n. Also let ∇ be the Levi-Civita connection of (M, g).

6.1 Semi-Riemannian Eikonal Equations

First let's formally state the definition of semi-Riemannian eikonal equations.

Definition 6.1.1 Let $f : (M, g) \to \mathbf{R}$ be a map. Then f is said to satisfy a *semi-Riemannian eikonal equation* on (M, g) if

$$g(\nabla f, \nabla f) = k_0,$$

where k_0 is either -1, 0 or 1.

Remark 6.1.1 Let $f : (M, g) \to \mathbf{R}$ be a map, where M is connected. Note that, if f satisfies $g(\nabla f(p), \nabla f(p)) = k_0$ at each $p \in M$, where k_0 is either -1, 0 or 1, then since $g(\nabla f, \nabla f)$ is continuous on M, f satisfies a semi-Riemannian eikonal equation on (M, g).

Definition 6.1.2 Let $f : (M, g) \to \mathbf{R}$ satisfy the semi-Riemannian eikonal equation $g(\nabla f, \nabla f) = k_0$ on (M, g).

(1) If $k_0 = -1$ then f is said to satisfy the *timelike eikonal equation on* (M, g).

(2) If $k_0 = 0$ then f is said to satisfy the *null eikonal equation on* (M, g).

(3) If $k_0 = 1$ then f is said to satisfy the *spacelike eikonal equation on* (M, g).

Note that, in general, a weakly semi-Riemannian map is not a semi-Riemannian map. But the solutions of semi-Riemannian eikonal equation (which are in fact weakly semi-Riemannian maps) are semi-Riemannian maps.

Proposition 6.1.1 *Let* $f : (M, g) \to \mathbf{R}$ *be a map. Then*

(1) $g(\nabla f, \nabla f) = k$ *if and only if* $f : (M, g) \to (\mathbf{R}, kdt \otimes dt)$ *is a semi-Riemannian submersion with fibers as semi-Riemannian submanifolds of* (M, g)*, where* k *is* -1 *or* 1.

(2) $g(\nabla f, \nabla f) = 0$ *if and only if* $f : (M, g) \to (\mathbf{R}, kdt \otimes dt)$ *is a semi-Riemannian map with vanishing nondegenerate rank, where* k *is either of* -1 *or* 1. *In particular, if* $g(\nabla f, \nabla f) = 0$ *with* $\nabla f(p) \neq 0$ *at each* $p \in M$ *then* $f : (M, g) \to (\mathbf{R}, kdt \otimes dt)$ *is a semi-Riemannian submersion with fibers as degenerate submanifolds of* (M, g).

Proof. (1): Let $f : (M, g) \to \mathbf{R}$ be a map. Then first note that, at each $p \in M$,

$$\begin{aligned}
&(kdt \otimes dt)(f_*(\nabla f(p)), f_*(\nabla f(p))) \\
&\quad = (kdt \otimes dt)(g(\nabla f(p), \nabla f(p))\frac{d}{dt} \circ f, g(\nabla f(p), \nabla f(p))\frac{d}{dt} \circ f) \\
&\quad = kg(\nabla f(p), \nabla f(p))^2.
\end{aligned}$$

Now observe that $g(\nabla f(p), \nabla f(p)) \neq 0$ at each $p \in M$ if and only if f is a submersion with fibers as semi-Riemannian hypersurfaces in (M, g). Hence in this case,

$$(\mathcal{H}(p), g_{p\mathcal{H}(p)}) = (\text{span } \{\nabla f(p)\}, g_{p\text{span } \{\nabla f(p)\}})$$

and

$$(A_2(f(p)), k(dt \otimes dt)_{f(p)\,A_2(f(p))}) = (T_{f(p)}\mathbf{R}, k(dt \otimes dt)_{f(p)})$$

are 1-dimensional inner product spaces at each $p \in M$. Then by the natural identification of $\overline{f_{*p}}$ and $f_{*p}\mid_{\mathcal{H}(p)}$ at each $p \in M$,

$$\overline{f_{*p}} = f_{*p}\mid_{\mathcal{H}(p)}: (\mathcal{H}(p), g_{p\mathcal{H}(p)}) \rightarrow (T_{f(p)}\mathbf{R}, k(dt \otimes dt)_{f(p)})$$

is an isometry if and only if $kg(\nabla f(p), \nabla f(p))^2 = g(\nabla f(p), \nabla f(p))$ if and only if $kg(\nabla f(p), \nabla f(p)) = 1$ if and only if $g(\nabla f(p), \nabla f(p)) = k$.

(2): Note that, if $g(\nabla f(p), \nabla f(p)) = 0$ for $p \in M$ then either $\nabla f(p) = 0$ or $\nabla f(p)$ is null. Now, if $\nabla f(p) = 0$ then $\mathcal{V}(p) = T_pM$, and hence $\mathcal{H}(p) = \{0\}$. Therefore

$$f_{*p}\mid_{\mathcal{H}(p)}: \mathcal{H}(p) = \{0\} \rightarrow A_2(f(p)) = \{0\}$$

is the zero linear transformation and hence $\overline{f_{*p}}$ is the zero isometry. That is f is semi-Riemannian with vanishing nondegenerate rank at $p \in M$. If $\nabla f(p)$ is null, then since $\mathcal{H}(p) = L_1(p)$ (see Proposition 1.1.4), $\overline{\mathcal{H}}(p) = \{0\}$. Therefore

$$\overline{f_{*p}} : \overline{\mathcal{H}}(p) = \{0\} \rightarrow \overline{A_2}(f(p))$$

is the zero linear transformation and hence the zero (into) isometry. That is, f is semi-Riemannian with vanishing nondegenerate rank at $p \in M$. Conversely if $f : (M, g) \rightarrow (\mathbf{R}, k(dt \otimes dt))$ is a semi-Riemannian map with vanishing nondegenerate rank then $\|f_*\|^2 = 0$. Thus as it was shown at the begining of this chapter that $kg(\nabla f, \nabla f) = 0$. Since k is either -1 or 1, it follows that $g(\nabla f, \nabla f) = 0$. In particular, note that if $\nabla f(p) \neq 0$ at each $p \in M$ then, f is a submersion. Hence, since it also follows from $g(\nabla f, \nabla f) = 0$ that f is a semi-Riemannian map, f is a semi-Riemannian submersion. Moreover since $L_1 = \mathcal{H} =$ span $\{\nabla f\}$, fibers of f are degenerate hypersurfaces in (M, g). □

A vector field Z on (M, g) is called *geodesic* if $\nabla_Z Z = 0$. Hence a vector field on (M, g) is geodesic if and only if the integral curves of Z are geodesics of (M, g). Indeed, if Z is a vector field on (M, g) and $\gamma : I \rightarrow M$ is an integral curve of Z then since

$$\nabla_{\frac{d}{dt}} \dot{\gamma} = \nabla_{\frac{d}{dt}} (Z \circ \gamma) = \nabla_{\dot{\gamma}} Z = \nabla_{Z \circ \gamma} Z = (\nabla_Z Z) \circ \gamma,$$

each integral curve γ of Z is a geodesic of (M, g) if and only if Z is a geodesic vector field on (M, g).

Proposition 6.1.2 *Let $f : (M, g) \rightarrow \mathbf{R}$ be a map. If f satisfies a semi-Riemannian eikonal equation $g(\nabla f, \nabla f) = k_0$ then ∇f is a geodesic vector field on (M, g) and is unit, provided that it is nowhere vanishing and nonnull.*

Proof. By Corollary 3.1.1,

$$0 = \nabla g(\nabla f, \nabla f) = 2\nabla_{\nabla f} \nabla f.$$

Thus ∇f is a geodesic vector field on (M, g). Hence if ∇f is nowhere vanishing and nonnull then $g(\nabla f, \nabla f) = k$, where k is either -1 or 1. □

As a corollary to the Proposition above, we have the following relation between the solutions of null eikonal equation and harmonic functions on 2-dimensional Lorentzian manifolds.

Corollary 6.1.1 *Let (M, g) be a 2-dimensional Lorentzian manifold and $f : (M, g) \to \mathbf{R}$ be a solution of the null eikonal equation on (M, g) with $\nabla f(p) \neq 0$ at each $p \in M$. Then f is a harmonic map on (M, g).*

Proof. First note that, by Proposition 6.1.2, $U_1 = \nabla f$ is a geodesic null vector field on (M, g). Also let U_2 be the null vector field on (M, g) determined by $g(U_1, U_2) = -\frac{1}{2}$. Then $Z_1 = U_1 + U_2$ and $Z_2 = U_1 - U_2$ form an orthonormal (global) frame. Hence,

$$\begin{aligned}
\Delta f &= -\operatorname{div} \nabla f \\
&= -\operatorname{div} U_1 \\
&= g(\nabla_{Z_1} U_1, Z_1) - g(\nabla_{Z_2} U_1, Z_2) \\
&= g(\nabla_{U_2} U_1, U_2) - g(\nabla_{U_2} U_1, U_2) \\
&= 0.
\end{aligned}$$

Thus f is harmonic on (M, g). □

In general, there is no known method to obtain even a local solution of the generalized eikonal equation. But for a certain special case, a solution of the generalized eikonal equation may be obtained by using solutions of semi-Riemannian eikonal equations.

Proposition 6.1.3 *Let $f = (f^1, \ldots, f^k) : (M, g) \to (\mathbf{R}^k, \cdot)$ be a map, where $\cdot$ is the usual Euclidean metric tensor $\cdot = \sum_{i=1}^{k} dx^i \otimes dx^i$ on $\mathbf{R}^k$. If $(\ker f_{*p}, g_{p \ker f_{*p}})$ is a nondegenerate subspace of (T_pM, g_p) at each $p \in M$, $\{\nabla f^1(p), \ldots, \nabla f^k(p)\}$ is a linearly independent set at each $p \in M$ and $g(\nabla f^i, \nabla f^i) = 1$, $i = 1, \ldots, k$, then f satisfies the generalized eikonal equation $\|f_*\|^2 = \operatorname{rank} \overline{f_*} = k$.*

Proof. Clearly, since $\{\nabla f^1(p), \ldots, \nabla f^k(p)\}$ is a linearly independent set at each $p \in M$,

$$f_{*p} = (df^1(p), \ldots, df^k(p)) : T_pM \to T_{f(p)}\mathbf{R}^k$$

is surjective at each $p \in M$, and hence rank $f = k$. Also since (ker f_{*p}, $g_{p\ker f_{*p}}$) is nondegenerate at each $p \in M$, rank $\overline{f_*}$ = rank $f = k$. To show that f satisfies the generalized eikonal equation $\|f_*\|^2$ = rank $\overline{f_*} = k$, let $\{X_1, \ldots, X_n\}$ be an orthonormal local frame. Then

$$\begin{aligned}
\|f_*\|^2 &= \sum_{i=1}^{n} g(X_i, X_i)((f_*X_i) \cdot (f_*X_i)) \\
&= \sum_{i=1}^{n} g(X_i, X_i)((df^1(X_i), \ldots, df^k(X_i)) \cdot (df^1(X_i), \ldots, df^k(X_i))) \\
&= \sum_{i=1}^{n} g(X_i, X_i)((df^1(X_i))^2 + \cdots + (df^k(X_i))^2) \\
&= \sum_{i=1}^{n} g(X_i, X_i)g(\nabla f^1, X_i)^2 + \cdots + \sum_{i=1}^{n} g(X_i, X_i)g(\nabla f^k, X_i)^2 \\
&= g(\nabla f^1, \nabla f^1) + \cdots + g(\nabla f^k, \nabla f^k) \\
&= k = \text{rank } \overline{f_*}.
\end{aligned}$$

□

Locally semi-Riemannian eikonal equations can be solved by using the Cauchy's method of characteristics, whereby, solutions to the partial differential equation $g(\nabla f, \nabla f) = k_0$ are obtained by solving the system of ordinary differential equations $\gamma'(t) = ((\nabla f) \circ \gamma)(t)$. But globally, a semi-Riemannian eikonal equation may not have a solution as we will show next.

Definition 6.1.3 Let $f : (M, g) \to \mathbf{R}$ be a map. Then f is said to satisfy

(1) *timelike eikonal inequality* if $g(\nabla f, \nabla f) < 0$,

(2) *spacelike eikonal inequality* if $g(\nabla f, \nabla f) > 0$.

Clearly timelike and spacelike eikonal inequalities are less restrictive than timelike and spacelike eikonal equations, respectively. But eikonal inequalities and equations are related up to a conformal metric tensor. Recall that two metric tensors g

and g_c on a manifold M are called *conformally related* if $g_c = \varphi g$, where $\varphi : M \to (0, \infty)$ is a map.

For future references, first we give the relations between some differential operators defined by conformally related metric tensors.

Lemma 6.1.1 *Let (M, g) be a semi-Riemannian manifold of dimension n and let $g_c = \varphi g$ be a conformally related metric tensor to g. Then*

(1) $\overset{c}{\nabla} = \frac{1}{\varphi}\nabla$, *where ∇ and $\overset{c}{\nabla}$ are the gradients on (M, g) and (M, g_c), respectively.*

(2) *For X, $Y \in \Gamma TM$,*

$$\overset{c}{\nabla}_X Y = \nabla_X Y + \frac{1}{2\varphi}X(\varphi)Y + \frac{1}{2\varphi}Y(\varphi)X - \frac{1}{2\varphi}g(X,Y)\nabla\varphi$$

where ∇ and $\overset{c}{\nabla}$ are the Levi-Civita connections of (M, g) and (M, g_c), respectively.

(3) *If $f : M \to \mathbf{R}$ is a map then, for X, $Y \in \Gamma TM$,*

$$\begin{aligned} H_f^c(X,Y) &= H_f(X,Y) - \tfrac{1}{2\varphi}[g(\nabla\varphi, X)g(\nabla f, Y) \\ &\quad + g(\nabla f, X)g(\nabla\varphi, Y) - g(\nabla\varphi, \nabla f)g(X,Y)] \end{aligned}$$

where H_f and H_f^c are the Hessian forms of f on (M, g) and (M, g_c), respectively.

(4) *If $f : M \to \mathbf{R}$ is a map, then*

$$\overset{c}{\Delta} f = \frac{1}{\varphi}\Delta f - \frac{n-2}{2\varphi}g(\nabla\varphi, \nabla f),$$

where Δ and $\overset{c}{\Delta}$ are the Laplacians on (M, g) and (M, g_c), respectively.

Proof. (1) Let $X \in \Gamma TM$ and $f : M \to \mathbf{R}$ be a map. Then since,

$$g_c(\overset{c}{\nabla} f, X) = X(f) = g(\nabla f, X) = \frac{1}{\varphi}g_c(\nabla f, X) = g_c(\frac{1}{\varphi}\nabla f, X),$$

it follows that $\overset{c}{\nabla} f = \frac{1}{\varphi}\nabla f$.

(2) By Koszul formula (see Theorem 2.2.2), for X, Y, $Z \in \Gamma TM$,

$$\begin{aligned} g_c(\overset{c}{\nabla}_X Y, Z) &= \tfrac{1}{2}\left[X g_c(Y,Z) + Y g_c(X,Z) - Z g_c(X,Y)\right. \\ &\quad \left. + g_c(Z,[X,Y]) + g_c(Y,[Z,X]) - g_c(X,[Y,Z])\right]. \end{aligned}$$

Thus

$$\begin{aligned} \varphi g(\overset{c}{\nabla}_X Y, Z) &= \tfrac{1}{2}\left[X(\varphi)g(Y,Z) + \varphi X g(Y,Z)\right. \\ &\quad + Y(\varphi)g(X,Z) + \varphi Y g(X,Z) \\ &\quad - Z(\varphi)g(X,Y) - \varphi Z g(X,Y) \\ &\quad \left. + \varphi g(Z,[X,Y]) + \varphi g(Y,[Z,X]) - \varphi g(X,[Y,Z])\right] \\ &= \varphi g(\nabla_X Y, Z) + \tfrac{1}{2}X(\varphi)g(Y,Z) \\ &\quad + \tfrac{1}{2}Y(\varphi)g(X,Z) - \tfrac{1}{2}g(X,Y)g(\nabla\varphi, Z) \\ &= \varphi g(\nabla_X Y, Z) \\ &\quad + g(\tfrac{1}{2}X(\varphi)Y + \tfrac{1}{2}Y(\varphi)X - \tfrac{1}{2}g(X,Y)\nabla\varphi, Z). \end{aligned}$$

Hence

$$\overset{c}{\nabla}_X Y = \nabla_X Y + \frac{1}{2\varphi}X(\varphi)Y + \frac{1}{2\varphi}Y(\varphi)X - \frac{1}{2\varphi}g(X,Y)\nabla\varphi.$$

(3) By (1) and (2), for X, $Y \in \Gamma TM$,

$$\begin{aligned} H_f^c(X,Y) &= g_c(\overset{c}{\nabla}_X \overset{c}{\nabla} f, Y) \\ &= g_c(\overset{c}{\nabla}_X (\tfrac{1}{\varphi}\nabla f), Y) \\ &= g_c(-\tfrac{X(\varphi)}{\varphi^2}\nabla f, Y) + g_c(\tfrac{1}{\varphi}\overset{c}{\nabla}_X \nabla f, Y) \\ &= -\tfrac{X(\varphi)}{\varphi}g(\nabla f, Y) + g(\nabla_X \nabla f, Y) \\ &\quad + \tfrac{X(\varphi)}{2\varphi}g(\nabla f, Y) + \tfrac{(\nabla f)(\varphi)}{2\varphi}g(X,Y) \\ &\quad - \tfrac{1}{2\varphi}g(X,\nabla f)g(\nabla f, Y) \\ &= H_f(X,Y) - \tfrac{1}{2\varphi}\left[g(\nabla\varphi, X)g(\nabla f, Y)\right. \\ &\quad \left. + g(\nabla f, X)g(\nabla\varphi, Y) - g(\nabla\varphi, \nabla f)g(X,Y)\right]. \end{aligned}$$

(4) Let $\{X_1, \ldots, X_n\}$ be an orthonormal local frame with respect to g. Then note that, $\{X_1^c = \frac{X_1}{\varphi^{1/2}}, \ldots, X_n^c = \frac{X_n}{\varphi^{1/2}}\}$ is an orthonormal local frame with respect to g_c. Hence by (3),

$$
\begin{aligned}
\overset{c}{\Delta} f &= -\sum_{i=1}^{n} g_c(X_i^c, X_i^c) H_f^c(X_i^c, X_i^c) \\
&= -\frac{1}{\varphi}\sum_{i=1}^{n} g(X_i, X_i) H_f^c(X_i, X_i) \\
&= -\frac{1}{\varphi}\sum_{i=1}^{n} g(X_i, X_i) H_f(X_i, X_i) \\
&\quad + \frac{1}{2\varphi^2}\sum_{i=1}^{n} g(X_i, X_i) g(\nabla\varphi, X_i) g(\nabla f, X_i) \\
&\quad + \frac{1}{2\varphi^2}\sum_{i=1}^{n} g(X_i, X_i) g(\nabla f, X_i) g(\nabla\varphi, X_i) \\
&\quad + \frac{1}{2\varphi^2} g(\nabla\varphi, \nabla f)\sum_{i=1}^{n} g(X_i, X_i) g(X_i, X_i) \\
&= \frac{1}{\varphi}\Delta f + \frac{1}{2\varphi^2} g(\nabla\varphi, \nabla f) + \frac{1}{2\varphi^2} g(\nabla\varphi, \nabla f) \\
&\quad - \frac{n}{2\varphi^2} g(\nabla\varphi, \nabla f) \\
&= \frac{1}{\varphi}\Delta f - \frac{n-2}{2\varphi^2} g(\nabla\varphi, \nabla f).
\end{aligned}
$$

□

Proposition 6.1.4 *Let $f : (M, g) \to \mathbf{R}$ be a map. Then, f satisfies the timelike (resp., spacelike) eikonal inequality on (M, g) if and only if f satisfies the timelike (resp., spacelike) eikonal equation on (M, g_c), where $g_c = |g(\nabla f, \nabla f)|\, g$.*

Proof. Recall that, if $g_c = \varphi g$, where $\varphi : M \to (0, \infty)$ is a map, then the gradient $\overset{c}{\nabla}$ on (M, g_c) is related to the gradient ∇ on (M, g) by $\overset{c}{\nabla} = \frac{1}{\varphi}\nabla$. (See Lemma 6.1.1-(1)).

Hence by setting $\varphi = |g(\nabla f, \nabla f)|$, if f satisfies $g(\nabla f, \nabla f) < 0$ (resp., $g(\nabla f, \nabla f) > 0$) then

$$\begin{aligned} g_c(\overset{c}{\nabla} f, \overset{c}{\nabla} f) &= |g(\nabla f, \nabla f)|\, g(\frac{1}{|g(\nabla f, \nabla f)|}\nabla f, \frac{1}{|g(\nabla f, \nabla f)|}\nabla f) \\ &= \frac{g(\nabla f, \nabla f)}{|g(\nabla f, \nabla f)|} = -1 \quad (\text{resp.}, = 1). \end{aligned}$$

Conversely, if f satisfies $g_c(\overset{c}{\nabla} f, \overset{c}{\nabla} f) = -1$ (resp., $g_c(\overset{c}{\nabla} f, \overset{c}{\nabla} f) = 1$) then

$$\begin{aligned} g(\nabla f, \nabla f) &= \frac{1}{|g(\nabla f, \nabla f)|} g_c(|g(\nabla f, \nabla f)|\overset{c}{\nabla} f, |g(\nabla f, \nabla f)|\overset{c}{\nabla} f) \\ &= |g(\nabla f, \nabla f)|\, g_c(\overset{c}{\nabla} f, \overset{c}{\nabla} f) \\ &= -|g(\nabla f, \nabla f)| < 0 \quad (\text{resp.}, = |g(\nabla f, \nabla f)| > 0). \end{aligned}$$

□

Remark 6.1.2 A Lorentzian manifold (M, g) is called *stably causal* if and only if timelike eikonal inequality on (M, g) has a solution. (See Proposition B.2.4). Note that stable causality is a conformally invariant property of (M, g). That is, if (M, g) is a stably causal Lorentzian manifold and $g_c = \varphi g$ is a conformally related metric tensor to g, then (M, g_c) is also stably causal. Indeed, if $g(\nabla f, \nabla f) < 0$ then $g_c(\overset{c}{\nabla} f, \overset{c}{\nabla} f) = \varphi g(\frac{1}{\varphi}\nabla f, \frac{1}{\varphi}\nabla f) = \frac{1}{\varphi} g(\nabla f, \nabla f) < 0$. Thus, for example, if a Lorentzian manifold (M, g) is not stably causal then timelike eikonal equation has no solution on any (M, g_c), where g_c is a conformally related metric tensor to g.

6.2 Affine Solutions of Semi-Riemannian Eikonal Equations

In general, an affine map $f : (M_1, g_1) \to (M_2, g_2)$ is not necessarily a semi-Riemannian map and conversely, a semi-Riemannian map $f : (M_1, g_1) \to (M_2, g_2)$ is not necessarily an affine map. However, we show in this section that a nonconstant affine map $f : (M, g) \to (\mathbf{R}, kdt \otimes dt)$, where k is either of -1 or 1, differs from a semi-Riemannian submersion only by a multiplicative constant. Later, we investigate the conditions under which a semi-Riemannian map $f : (M, g) \to (\mathbf{R}, kdt \otimes dt)$, where k is either of -1 or 1, is an affine map. For these, first we introduce the concept of an affine map $f : (M, g) \to \mathbf{R}$. To understand the origin of this concept, let's first compute the second fundamental form of a map $f : (M, g) \to (\mathbf{R}, kdt \otimes dt)$, where k is either of -1 or 1. For this, let D be the Levi-Civita connection of $(\mathbf{R}, kdt \otimes dt)$.

(Note that, $(\mathbf{R}, -dt \otimes dt)$ and $(\mathbf{R}, dt \otimes dt)$ have the same Levi-Civita connection). Then for $X, Y \in \Gamma TM$,

$$\begin{aligned}
(\nabla f_*)(X,Y) &= D_X f_* Y - f_*(\nabla_X Y) \\
&= D_X(g(\nabla f, Y)\tfrac{d}{dt} \circ f) - g(\nabla f, \nabla_X Y)\tfrac{d}{dt} \circ f \\
&= Xg(\nabla f, Y)\tfrac{d}{dt} \circ f + g(\nabla f, Y) D_X(\tfrac{d}{dt} \circ f) - g(\nabla f, \nabla_X Y)\tfrac{d}{dt} \circ f \\
&= g(\nabla_X \nabla f, Y)\tfrac{d}{dt} \circ f + g(\nabla f, Y) D_{f_* X}\tfrac{d}{dt} \\
&= H_f(X,Y)\tfrac{d}{dt} \circ f + g(\nabla f, Y) g(\nabla f, X) D_{\frac{d}{dt} \circ f}\tfrac{d}{dt} \\
&= H_f(X,Y)\tfrac{d}{dt} \circ f,
\end{aligned}$$

since $D_{\frac{d}{dt} \circ f}\frac{d}{dt} = (D_{\frac{d}{dt}}\frac{d}{dt}) \circ f = 0$. Hence $f : (M,g) \to (\mathbf{R}, kdt \otimes dt)$ is affine if and only if $H_f = 0$ if and only if $h_f = 0$. This motivates us to make the following definition:

Definition 6.2.1 Let $f : (M,g) \to \mathbf{R}$ be a map. Then f is called *affine* if $H_f = 0$, or equivalently, $h_f = 0$.

Also recall that, if $I \subseteq \mathbf{R}$ is an open interval, then a function $f : I \to \mathbf{R}$ is called *affine* if $f(t) = At + B$, where $A, B \in \mathbf{R}$.

Proposition 6.2.1 *Let $f : (M,g) \to \mathbf{R}$ be a map. Then the following are equivalent:*

(1) *f is an affine map.*

(2) *If $\gamma : I \to M$ is a geodesic of (M,g) then $f \circ \gamma : I \to \mathbf{R}$ is an affine function.*

(3) *∇f is a parallel vector field on (M,g).*

Proof. (1)$\Longleftrightarrow$(2): Let $\gamma : I \to M$ be a geodesic of (M,g). Then since

$$\frac{d}{dt}(f \circ \gamma) = g((\nabla f) \circ \gamma, \dot{\gamma}),$$

we obtain

$$\begin{aligned}\frac{d^2}{dt^2}(f\circ\gamma) &= \frac{d}{dt}g((\nabla f)\circ\gamma,\dot\gamma)\\ &= g(\nabla_{\frac{d}{dt}}((\nabla f)\circ\gamma),\dot\gamma)\\ &= g(\nabla_{\dot\gamma}\nabla f,\dot\gamma)\\ &= H_f(\dot\gamma,\dot\gamma).\end{aligned}$$

Thus $f\circ\gamma : I \to \mathbf{R}$ is an affine function if and only if $H_f(\dot\gamma,\dot\gamma) = 0$. Hence, since H_f is a symmetric $(0,2)$-tensor field on M, $f\circ\gamma : I \to \mathbf{R}$ is an affine function for each geodesic $\gamma : I \to M$ of (M,g) if and only if $H_f = 0$, that is, $f : (M,g) \to \mathbf{R}$ is an affine map.

(1)$\Longleftrightarrow$(3): Since for every $X, Y \in \Gamma TM$,

$$H_f(X,Y) = g(\nabla_X\nabla f, Y),$$

it follows that $H_f = 0$ if and only if $\nabla\nabla f = 0$, that is, ∇f is parallel. □

Remark 6.2.1 Note that since the geodesics of $(\mathbf{R}, kdt\otimes dt)$ are of the form $\gamma(t) = At + B$, where $A, B \in \mathbf{R}$, Proposition 6.2.1–(2) is equivalent to that f sends geodesics to geodesics. (Also see Proposition 3.2.2).

Proposition 6.2.2 *Let $f : (M,g) \to \mathbf{R}$ be an affine map, where M is connected. Then,*

(1) rank *of f is constant on M.*

(2) *$g(\nabla f, \nabla f)$ is constant on M.*

Proof. (1): First note that the rank of f is constant on M if and only if either $\nabla f(p) = 0$ at every $p \in M$ or $\nabla f(p) \neq 0$ at every $p \in M$. Let $p, q \in M$ and $\gamma : [a,b] \to M$ be a curve with $\gamma(a) = p$, $\gamma(b) = q$. Then by Proposition 6.1.3–(3), since ∇f is a parallel vector field on (M,g), $(\nabla f)\circ\gamma$ is a parallel vector field along γ. By the uniqueness of parallel vector fields along a curve with respect to initial condition, $\nabla f(p) = 0$ if and only if $\nabla f(q) = 0$, and $\nabla f(p) \neq 0$ if and only if $\nabla f(q) \neq 0$. Thus, either $\nabla f(p) = 0$ for every $p \in M$ or $\nabla f(p) \neq 0$ for every $p \in M$.

(2): By Corollary 3.1.1 and Proposition 6.2.1–(3),

$$\nabla g(\nabla f, \nabla f) = 2\nabla_{\nabla f}\nabla f = 0.$$

Thus $g(\nabla f, \nabla f)$ is constant on M. □

Proposition 6.2.3 *Let $f : (M, g) \to \mathbf{R}$ be a nonconstant affine map, where M is connected. Then the map $h : (M, g) \to (\mathbf{R}, kdt \otimes dt)$ defined by*

$$h = \begin{cases} \dfrac{1}{|g(\nabla f, \nabla f)|^{1/2}} f & \text{if } g(\nabla f, \nabla f) \neq 0 \\ f & \text{if } g(\nabla f, \nabla f) = 0 \end{cases},$$

where

$$k = \begin{cases} -1 & \text{if } g(\nabla f, \nabla f) < 0 \\ -1 \text{ or } 1 & \text{if } g(\nabla f, \nabla f) = 0 \\ 1 & \text{if } g(\nabla f, \nabla f) > 0 \end{cases},$$

is a semi-Riemannian submersion with totally geodesic fibers. Hence in particular, h satisfies a semi-Riemannian eikonal equation.

Proof. Note that, by Proposition 6.2.2–(1), since f is a nonconstant affine map, $\nabla f(p) \neq 0$ at every $p \in M$. Also by Proposition 6.2.2–(2), since $g(\nabla f, \nabla f)$ is constant on M, the map $h : (M, g) \to (\mathbf{R}, kdt \otimes dt)$ defined by

$$h = \begin{cases} \dfrac{1}{|g(\nabla f, \nabla f)|^{1/2}} f & \text{if } g(\nabla f, \nabla f) \neq 0 \\ f & \text{if } g(\nabla f, \nabla f) = 0 \end{cases}$$

satisfies a semi-Riemannian eikonal equation $g(\nabla h, \nabla h) = k_0$. Hence by Proposition 6.1.1, $h : (M, g) \to (\mathbf{R}, kdt \otimes dt)$ is a semi-Riemannian submersion, where

$$k = \begin{cases} -1 & \text{if } g(\nabla f, \nabla f) < 0 \\ -1 \text{ or } 1 & \text{if } g(\nabla f, \nabla f) = 0 \\ 1 & \text{if } g(\nabla f, \nabla f) > 0 \end{cases}.$$

Moreover, fibers of h are totally geodesic submanifolds of (M, g). Indeed, since $h : (M, g) \to \mathbf{R}$ is also an affine map, for every $X, Y \in \Gamma \ker h_*$,

$$g(\nabla_X Y, \nabla h) = Xg(Y, \nabla h) - g(Y, \nabla_X \nabla h) = 0.$$

Hence $\nabla_X Y \perp \nabla h$, that is, $\nabla_X Y \in \Gamma \ker h_*$. Thus fibers of h are totally geodesic. □

Now we investigate some sufficient conditions under which a solution of a semi-Riemannian eikonal equation is an affine map.

Proposition 6.2.4 *Let $f : (M, g) \to (\mathbf{R}, k dt \otimes dt)$ be a semi-Riemannian submersion, where k is either of -1 or 1. If the fibers of f are totally geodesic then $f : (M, g) \to \mathbf{R}$ is an affine map.*

Proof. Let X, $Y \in \Gamma \ker f_*$. Then since $\nabla_X Y \in \Gamma \ker f_*$,

$$\begin{aligned} 0 = g(\nabla_X Y, \nabla f) &= Xg(Y, \nabla f) - g(Y, \nabla_X \nabla f) \\ &= -H_f(X, Y). \end{aligned}$$

Also, since ∇f is a geodesic vector field on (M, g) by Proposition 6.1.2, for $X \in \Gamma TM$,

$$H_f(X, \nabla f) = H_f(\nabla f, X) = g(\nabla_{\nabla f} \nabla f, X) = 0.$$

Thus, $H_f = 0$, that is, f is an affine map. □

Proposition 6.2.5 *Let $f : (M, g) \to \mathbf{R}$ be a map, where M is connected. If*

(1) *h_f is diagonalizable at each $p \in M$,*

(2) *f satisfies a semi-Riemannian eikonal equation $g(\nabla f, \nabla f) = k_0$, (that is, $f : (M, g) \to (\mathbf{R}, k dt \otimes dt)$ is a semi-Riemannian map, where k is either of -1 or 1),*

(3) *$Ric(\nabla f, \nabla f) \geq 0$,*

(4) *∇f is a complete vector field on M,*

then f is an affine map.

Proof. Note that, since f satisfies a semi-Riemannian eikonal equation, by Corollary 3.4.1,

$$0 = \|h_f\|^2 + Ric(\nabla f, \nabla f) - g(\nabla \Delta f, \nabla f).$$

Thus it follows from $Ric(\nabla f, \nabla f) \geq 0$ and h_f is diagonalizable at each $p \in M$ (see Lemma 1.2.1) that

$$g(\nabla \Delta f, \nabla f) \geq \|h_f\|^2 \geq \frac{1}{n}(\text{trace } h_f)^2 = \frac{1}{n}(\Delta f)^2.$$

Now we show that $\Delta f = 0$ on (M, g). For let $\gamma : \mathbf{R} \to M$ be an integral curve of ∇f. Then

$$((\Delta f)\circ\gamma)'(t) = g((\nabla\Delta f)\circ\gamma,(\nabla f)\circ\gamma)(t) \geq \frac{1}{n}((\Delta f)\circ\gamma)^2(t) \geq 0.$$

Thus $((\Delta f)\circ\gamma)(t)$ is a nondecreasing function on $\mathbf{R}$. Suppose $((\Delta f)\circ\gamma)(t_0) \neq 0$ and assume $((\Delta f)\circ\gamma)(t_0) > 0$. (The case $((\Delta f)\circ\gamma)(t_0) < 0$ is similar). Then $((\Delta f)\circ\gamma)(t) > 0$ for all $t \geq t_0$ and it follows from

$$((\Delta f)\circ\gamma)'(t) \geq \frac{1}{n}((\Delta f)\circ\gamma)^2(t)$$

that

$$\int_{t_0}^{t} \frac{((\Delta f)\circ\gamma)'(s)}{((\Delta f)\circ\gamma)^2(s)}ds \geq \frac{1}{n}\int_{t_0}^{t} ds$$

for all $t \geq t_0$. Thus

$$\frac{1}{((\Delta f)\circ\gamma)(t_0)} - \frac{1}{n}(t-t_0) \geq \frac{1}{((\Delta f)\circ\gamma)(t)}$$

for all $t \geq t_0$. But this yields a contradiction as $t \to \infty$. Thus $\Delta f = 0$ on M. Then it follows that $\|h_f\|^2 = 0$ and hence, $h_f = 0$ since h_f is diagonalizable at each $p \in M$. That is, f is affine. □

Remark 6.2.2 Here it is worth emphasizing a special case of Proposition 6.2.5 when (M, g) is a Riemannian manifold. Note that if (M, g) is taken to be a Riemannian manifold in Proposition 6.2.5 then, since h_f is diagonalizable at each $p \in M$, the assumption (a) of Proposition 6.2.5 can be dropped.

Now by using Proposition 6.2.5, we can characterize a certain subset of semi-Riemannian maps from an n-dimensional semi-Euclidean space to 1-dimensional semi-Euclidean space. Let $\cdot$ both denote the semi-Euclidean inner product of index ν and semi-Euclidean metric tensor of index ν on $\mathbf{R}^n$, that is, $\cdot\ = -\sum_{i=1}^{\nu} dx^i \otimes dx^i + \sum_{i=\nu+1}^{n} dx^i \otimes dx^i$. We call $(\mathbf{R}^n, \cdot)$ the *semi-Euclidean space of index* ν.

Proposition 6.2.6 *Let* $(\mathbf{R}^n, \cdot)$ *be the semi-Euclidean space of index* ν. *If* $f : (\mathbf{R}^n, \cdot) \to \mathbf{R}$ *is a solution of the semi-Riemannian eikonal equation* $\nabla f \cdot \nabla f = k_0$ *with diagonalizable Hessian tensor* h_f *at each* $p \in \mathbf{R}^n$ *then* f *is of the form* $f(p) = u \cdot p + b$, *where* $u \in \mathbf{R}^n$ *with* $u \cdot u = k_0$ *and* $b \in \mathbf{R}$.

Proof. Note that, by Proposition 6.1.2, ∇f is a geodesic vector field on $(\mathbf{R}^n, \cdot)$ and hence, since $(\mathbf{R}^n, \cdot)$ is geodesically complete, ∇f is a complete vector field on $\mathbf{R}^n$.

Then by Proposition 6.2.5, $f : (\mathbf{R}^n, \cdot) \to \mathbf{R}$ is an affine map. Thus since $H_f = 0$, f is of the form $f(p) = u \cdot p + b$, where $u \in \mathbf{R}^n$ and $b \in \mathbf{R}$. Also since $\nabla f \cdot \nabla f = k_0$, $u \cdot u = k_0$. □

Remark 6.2.3 Here it is worth emphasizing a special case of Proposition 6.2.6 when $(\mathbf{R}^n, \cdot)$ is the Euclidean space. Note that if $(\mathbf{R}^n, \cdot)$ is taken to be the Euclidean space (that is, the semi-Euclidean space of index 0) then, since h_f is diagonalizable at each $p \in \mathbf{R}^n$, the assumption that h_f is diagonalizable at each $p \in M$ can be dropped from the statement of Proposition 6.2.6. Hence in this case, Proposition 6.2.6 completely characterizes the Riemannian maps from n-dimensional Euclidean space to 1-dimensional Euclidean space. It is important to note that, even in the Euclidean case, Proposition 6.2.6 fails to hold if we remove a point from $\mathbf{R}^n$. For example, $f : (\mathbf{R}^n - \{p_0\}, \cdot) \to \mathbf{R}$ defined by

$$f(p) = ((p - p_0) \cdot (p - p_0))^{1/2} = \|p - p_0\|$$

is a solution of $\nabla f \cdot \nabla f = 1$ but f is not of the form $f(p) = u \cdot p + b$, where $\cdot$ is the Euclidean inner product on $\mathbf{R}^n - \{p_0\}$.

6.3 The Semi-Riemannian Regular Interval Theorem

The semi-Riemannian eikonal equations $g(\nabla f, \nabla f) = k$ on (M, g), where k is -1 or 1, are called *nonnull eikonal equations* on (M, g). As a consequence of the existence of a solution of a nonnull eikonal equation on (M, g), (M, g) may split into a parametrized semi-Riemannian manifold, which we call the *semi-Riemannian regular interval theorem.* In this section, we obtain the semi-Riemannian regular interval theorem and its generalization for semi-Riemannian manifolds. Also at the end of this section, we state one of its Riemannian versions for complete Riemannian manifolds.

First we introduce the parametrized semi-Riemannian product manifolds to fix our notation and terminology. Let (a, b) be an open interval in $\mathbf{R}$ and N be an n-dimensional manifold. Also let g_t be a (smooth) 1-parameter family of metric tensors of index ν_t on N parametrized over (a, b), that is, g_t is a metric tensor of index ν_t on N for each $t \in (a, b)$ and $t \mapsto g_{t_q}$ is a (smooth) curve in $T_q^*N \otimes T_q^*N$ at each $q \in N$. As one can immediately feel that the smoothness of the curve $t \mapsto g_{t_q}$ at each $q \in N$ makes g_t necessarily of the same index for every $t \in (a, b)$. To see this, let $\{x_1, \ldots, x_{\nu_{t_0}}, x_{\nu_{t_0}+1}, \ldots, x_n\}$ be an orthonormal basis for $(T_qN, g_{t_{0q}})$ for $t_0 \in (a, b)$ such that

$$g_{t_0q}(x_i, x_i) = \begin{cases} -1 & \text{for} \quad 1 \leq i \leq \nu_{t_0} \\ \\ 1 & \text{for} \quad \nu_{t_0+1} \leq i \leq n \end{cases},$$

where ν_{t_0} is the index of g_{t_0q}. Let

$$W_{-q} = \text{ span } \{x_1, \ldots, x_{\nu_{t_0}}\} \quad \text{and} \quad W_{+q} = \text{ span } \{x_{\nu_{t_0+1}}, \ldots, x_n\}.$$

We show that there exists $\delta > 0$ such that the restrictions of g_{t_q} to W_{-q} and W_{+q} are negative and positive definite, respectively, for all $t \in (t_0 - \delta, t_0 + \delta)$. Then by Remark 1.1.1, $\nu_t \geq \nu_{t_0}$ and $n - \nu_t \geq n - \nu_{t_0}$ for all $t \in (t_0 - \delta, t_0 + \delta)$ and hence $\nu_t = \nu_{t_0}$ for all $t \in (t_0 - \delta, t_0 + \delta)$. Hence it follows from the connectedness of (a, b) that g_{t_q} is of constant index on (a, b). Now we show that the restriction of g_{t_q} to W_{+q} is positive definite for all $t \in (t_0 - \delta_+, t_0 + \delta_+)$ for some $\delta_+ > 0$. (Showing that the restriction of g_{t_q} to W_{-q} is negative definite for all $t \in (t_0 - \delta_-, t_0 + \delta_-)$ for some δ_- > 0 is similar). Let

$$g_{ij}(t) = g_{t_q}(x_i, x_j) \quad \text{for} \quad 1 \leq i, j \leq n.$$

Since g_{ij} are continuous functions on (a, b), for $\varepsilon_+ = \frac{1}{n - \nu_{t_0} + 2}$, there exists $\delta_+ > 0$ such that

$$|\, g_{ii}(t) - 1 \,| < \varepsilon_+ \quad \text{for } \nu_{t_0} + 1 \leq i \leq n$$

and

$$|\, g_{ij}(t) \,| < \varepsilon_+ \quad \text{for } i \neq j, \quad \nu_{t_0} + 1 \leq i, j \leq n$$

whenever $|\, t - t_0 \,| < \delta_+$. Then for any $0 \neq x = \sum_{i=\nu_{t_0}+1}^{n} \lambda_i x_i \in W_{+q}$ it can be shown as in the proof of Lemma 2.1.1 that

$$g_{t_q}(x, x) = \sum_{i,j=\nu_{t_0}+1}^{n} \lambda_i \lambda_j g_{ij}(t) > 0$$

for all $t \in (t_0 - \delta_+, t_0 + \delta_+)$. Hence if $\delta = \min \{\delta_-, \delta_+\}$ then g_{t_q} is of constant index for all $t \in (t_0 - \delta, t_0 + \delta)$.

After this observation, the following definition will be meaningful:

Definition 6.3.1 Let (a, b) be an open interval in $\mathbf{R}$ and let N be a manifold with a (smooth) 1-parameter family of semi-Riemannian metric tensors g_t on N parametrized over (a, b). Then the *parametrized semi-Riemannian product* M_k of

(a, b) and N is defined to be the product manifold $M_k = (a, b) \times N$ with metric tensor $g_k = kdt \otimes dt \oplus g_t$, where k is either of -1 or 1. Here for $x, y \in T_{(t,q)}M_k$,

$$g_k(x, y) = kdt \otimes dt(\pi_{1*}x, \pi_{1*}y) + g_t(\pi_{2*}x, \pi_{2*}y),$$

where π_1 and π_2 are the projections of M_k onto (a, b) and N, respectively.

Remark 6.3.1 Note that, since g_t is of constant index for all t by the observation above, it is straightforward to check that g_k is a metric tensor on M_k.

Remark 6.3.2 Let $(M_k, g_k) = ((a, b) \times N, kdt \otimes dt \oplus g_t)$ be a parametrized semi-Riemannian product manifold. Then the following special cases are also of certain interest in semi-Riemannian geometry.

(a) If $g_{t_q} = \lambda(t, q)g_{N_q}$, where $\lambda : (a, b) \times N \to (0, \infty)$ and g_N is a fixed metric tensor on N, then (M_k, g_k) is called the *semi-Riemannian twisted product of* $((a, b), kdt \otimes dt)$ *and* (N, g_N) *with twisting function* λ.

(b) If $g_t = \lambda(t)g_N$, where $\lambda : (a, b) \to (0, \infty)$ and g_N is a fixed metric tensor on N, then (M_k, g_k) is called the *semi-Riemannian warped product of* $((a, b), kdt \otimes dt)$ *and* (N, g_N) *with warping function* λ.

(c) If $g_t = g_N$, where g_N is a fixed metric tensor on N, then (M_k, g_k) is called the *semi-Riemannian product of* $((a, b), kdt \otimes dt)$ *and* (N, g_N).

We also need the following lemma in the proof of the semi-Riemannian regular interval theorem.

Let Z be a complete vector field on a manifold M and $\Phi : \mathbf{R} \times M \to M$ be its flow. Hence if we define $\Phi_t(p) = \Phi(t, p)$, then $\Phi_t(p) = c_p(t)$, where $c_p(t)$ is the integral curve of Z with $c_p(0) = p$. Also note that $\Phi_t : M \to M$ is a 1-parameter group of diffeomorphisms of M parametrized over $\mathbf{R}$.

Lemma 6.3.1 [F2] *Let Z be a complete vector field on a manifold M and let Φ : $\mathbf{R} \times M \to M$ be its flow. If $(a, v) \in T_{(t,p)}(\mathbf{R} \times M)$ then*

$$\Phi_{*(t,p)}(a, v) = a(Z \circ \Phi_t)(p) + (\Phi_t)_{*p}v,$$

where $(Z \circ \Phi_t)(p) = Z_{\Phi_t(p)} = Z_{c_p(t)}$.

Proof. Let $(a, 0) \in T_{(t,p)}(\mathbf{R} \times M)$ and let $\gamma : \mathbf{R} \to \mathbf{R} \times M$ be $\gamma(s) = (t + as, p)$. (Note that $\gamma(0) = (t, p)$ and $\dot{\gamma}(0) = (a, 0)$). Then if $\chi : M \to \mathbf{R}$,

$$\begin{aligned}
(\Phi_{*(t,p)}(a,0))\chi &= (a,0)(\chi \circ \Phi) \\
&= \frac{d}{ds}(\chi \circ \Phi \circ \gamma) \mid_{s=0} \\
&= \frac{d}{ds}(\chi(\Phi(t+as,p))) \mid_{s=0} \\
&= \frac{d}{ds}(\chi(c_p(t+as))) \mid_{s=0} \\
&= a d\chi_{c_p(t)}(\dot{c_p}\,(t)) \\
&= a d\chi_{c_p(t)}(Z_{c_p(t)}) \\
&= a d\chi_{c_p(t)}((Z \circ \Phi_t)(p)) \\
&= a((Z \circ \Phi_t)(p))\chi.
\end{aligned}$$

Thus

$$\Phi_{*(t,p)}(a,0) = a(Z \circ \Phi_t)(p).$$

Also let $(0,v) \in T_{(t,p)}(\mathbf{R} \times M)$ and $\gamma : \mathbf{R} \to \mathbf{R} \times M$ be $\gamma(s) = (t, \alpha(s))$, where $\alpha : \mathbf{R} \to M$ with $\alpha(0) = p$ and $\dot{\alpha}\,(0) = v$. (Note that, $\gamma(0) = (t,p)$ and $\dot{\gamma}\,(0) = (0,v)$). Then if $\chi : M \to \mathbf{R}$,

$$\begin{aligned}
(\Phi_{*(t,p)}(0,v))\chi &= (0,v)(\chi \circ \Phi) \\
&= \frac{d}{ds}(\chi \circ \Phi \circ \gamma) \mid_{s=0} \\
&= \frac{d}{ds}(\chi(\Phi(t,\alpha(s)))) \mid_{s=0} \\
&= \frac{d}{ds}(\chi(\Phi_t(\alpha(s)))) \mid_{s=0} \\
&= d\chi_{\Phi_t(p)}((\Phi_t)_{*_p}\,\dot{\alpha}\,(0)) \\
&= d\chi_{\Phi_t(p)}((\Phi_t)_{*_p}v) \\
&= ((\Phi_t)_{*_p}v)\chi.
\end{aligned}$$

Hence $\Phi_{*(t,p)}(0,v) = (\Phi_t)_{*_p}v$. Thus it follows that

$$\begin{aligned}\Phi_{*(t,p)}(a,v) &= \Phi_{*(t,p)}((a,0)+(0,v)) \\ &= \Phi_{*(t,p)}(a,0)+\Phi_{*(t,p)}(0,v) \\ &= a(Z\circ\Phi_t)(p)+(\Phi_t)_{*_p}v\end{aligned}$$

□

Now we are ready to prove the semi-Riemannian regular interval theorem.

A map $f:(M_1,g_1)\to(M_2,g_2)$ is called a *conformal diffeomorphism* if $f:M_1\to M_2$ is a diffeomorphism and $f^*g_2=\varphi g_1$, where $\varphi:M_1\to(0,\infty)$. A conformal diffeomorphism $f:(M_1,g_1)\to(M_2,g_2)$ with $\varphi=1$ is called an *isometry*. Also (M_1,g_1) and (M_2,g_2) are called *conformally diffeomorphic* (resp., *isometric*) if there is a conformal diffeomorphism (resp., isometry) $f:(M_1,g_1)\to(M_2,g_2)$.

Theorem 6.3.1 (The Semi-Riemannian Regular Interval Theorem) *Let $f:(M,g)\to\mathbf{R}$ be a map, where M is connected. If f satisfies a nonnull eikonal equation $g(\nabla f,\nabla f)=k$ on (M,g) and ∇f is a complete vector field on M, then (M,g) is isometric to a parametrized semi-Riemannian product manifold $(M_k,g_k)=(\mathbf{R}\times N, kdt\otimes dt\oplus g_t)$, where k is either of -1 or 1. If moreover (still assumming that $g(\nabla f,\nabla f)=k$ and ∇f is complete) f is affine on (M,g) then (M,g) is isometric to a semi-Riemannian product manifold $(M_k,g_k)=(\mathbf{R}\times N, kdt\otimes dt\oplus g_N)$.*

Proof. By Proposition 6.1.2, ∇f is a complete nonnull geodesic vector field on M. Let $\Phi:\mathbf{R}\times M\to M$ be the flow of ∇f and $c_p(t)=\Phi_t(p)$ be an inextendable integral curve of ∇f (with $c_p(0)=p$). Then

$$\begin{aligned}\frac{d}{dt}f(c_p(t)) &= df_{c_p(t)}(\dot c(t)) \\ &= df_{\Phi_t(p)}(((\nabla f)\circ\Phi_t)(p)) \\ &= (df(\nabla f))_{\Phi_t(p)} \\ &= g(\nabla f,\nabla f)_{\Phi_t(p)}=k.\end{aligned}$$

Thus, for t, $t_0\in\mathbf{R}$,

$$f(c_p(t))-f(c_p(t_0))=k(t-t_0),$$

and in particular, for $t_0=0$,

$$f(c_p(t))=kt+f(p).$$

Hence, since c_p is complete, $f : (M, g) \to \mathbf{R}$ is surjective and it follows from Proposition 6.1.1–(1) that $f : (M, g) \to (\mathbf{R}, kdt \otimes dt)$ is a surjective semi-Riemannian submersion with fibers as semi-Riemannian submanifolds of (M, g). Also note that, if $p \in f^{-1}(s)$ then, since

$$f(\Phi_t(p)) = f(c_p(t)) = kt + f(p) = kt + s,$$

Φ_t maps the level set $f^{-1}(s)$ onto level set $f^{-1}(kt+s)$ diffeomorphically. Now, if we set

$$\Phi^0 = \Phi\,|_{\mathbf{R}\times f^{-1}(0)}: \mathbf{R} \times f^{-1}(0) \to M,$$

then since

$$\Phi^0(t, f^{-1}(0)) = \Phi(t, f^{-1}(0)) = \Phi_t(f^{-1}(0)) = f^{-1}(kt) \subseteq M,$$

Φ^0 maps $\{t\} \times f^{-1}(0)$ onto $f^{-1}(kt)$ and thus, Φ^0 is a bijective map. Now since Φ^0 is smooth and bijective, to show that it is a diffeomorphism, it suffices to show that Φ^0 is a local diffeomorphism. For this, first note that since Φ_t maps the level set $f^{-1}(s)$ onto the level set $f^{-1}(kt+s)$ diffeomorphically, if $p \in f^{-1}(s)$ then

$$\begin{aligned} (\Phi_t)_{*_p}(\ker f_{*p}) &= (\Phi_t)_{*_p}(T_p(f^{-1}(s))) \\ &= T_{\Phi_t(p)}(f^{-1}(kt+s)) \\ &= \ker f_{*\Phi_t(p)}, \end{aligned}$$

that is, $(\Phi_t)_{*_p}$ maps $\ker f_{*p}$ to $\ker f_{*\Phi_t(p)}$ isomorphically. Now let $(a, v) \in T_{(t,p)}(\mathbf{R} \times f^{-1}(0)) = \mathbf{R} \times \ker f_{*p}$. Then by Lemma 6.3.1,

$$\begin{aligned} \Phi^0_{*(t,p)}(a, v) &= \Phi_{*(t,p)}(a, v) \\ &= a((\nabla f) \circ \Phi_t)(p) + (\Phi_t)_{*_p} v. \end{aligned}$$

Hence, since $0 \neq \nabla f(\Phi_t(p))$ and $(\Phi_t)_{*_p} v \in \ker f_{*\Phi_t(p)}$ are orthogonal, if $\Phi^0_{*(t,p)}(a, v) = 0$ then it follows that $(a, v) = (0, 0)$, that is, $\Phi^0_{*(t,p)}$ is nonsingular and hence Φ^0 is a local diffeomorphism. Therefore,

$$\Phi^0 : \mathbf{R} \times f^{-1}(0) \to M$$

is a diffeomorphism. Now define a 1-parameter family of metric tensors g_t on $f^{-1}(0)$ parametrized over $\mathbf{R}$ by

$$g_t = (\Phi_t^* g)_0,$$

where $(\Phi_t^* g)_0$ is the restriction of $\Phi_t^* g$ to $f^{-1}(0)$. In other words,

$$g_t = (\Phi_t^* g)_0 = \Phi_t^*(g_{f^{-1}(kt)}),$$

where $g_{f^{-1}(kt)}$ is the restriction of g to $f^{-1}(kt)$.

Now we show that Φ^0 is an isometry between the parametrized semi-Riemannian product $(\mathbf{R} \times f^{-1}(0), kdt \otimes dt \oplus g_t)$ and (M, g). For this, let (a, u), $(b, v) \in T_{(t,p)}(\mathbf{R} \times f^{-1}(0))$. Then by Lemma 6.3.1,

$$\begin{aligned}
((\Phi^0)^* g)((a,u),(b,v)) &= g(\Phi^0_{*(t,p)}(a,u), \Phi^0_{*(t,p)}(b,v)) \\
&= g(\Phi_{*(t,p)}(a,u), \Phi_{*(t,p)}(b,v)) \\
&= g(a\nabla f(\Phi_t(p)) + (\Phi_t)_{*_p} u, b\nabla f(\Phi_t(p)) + (\Phi_t)_{*_p} v) \\
&= kab + (\Phi_t^* g)(u,v) \\
&= kab + (\Phi_t^* g)_0(u,v) \\
&= (kdt \otimes dt \oplus g_t)((a,u),(b,v)).
\end{aligned}$$

Hence $(\mathbf{R} \times f^{-1}(0), kdt \otimes dt \oplus g_t)$ is isometric to (M, g). Now by setting $N = f^{-1}(0)$, the claim follows.

Moreover, if f is an affine map as well, to show that $g_t = g_0$ for all $t \in \mathbf{R}$, it suffices to show that

$$\Phi_t \mid_{f^{-1}(0)} : (f^{-1}(0), g_0) \to (f^{-1}(kt), g_{f^{-1}(kt)})$$

is an isometry for each $t \in \mathbf{R}$. For this, let $X_0, Y_0 \in \Gamma T f^{-1}(0)$ and let $X, Y \in \Gamma TM$ such that $X_{\Phi_t(p)} = (\Phi_t)_* X_0$, $Y_{\Phi_t(p)} = (\Phi_t)_* Y_0$. Then since $[\nabla f, X] = 0 = [\nabla f, Y]$,

$$\nabla_{\nabla f} X = \nabla_X \nabla f = h_f(X) \quad \text{and} \quad \nabla_{\nabla f} Y = \nabla_Y \nabla f = h_f(Y).$$

Now since f is affine, $H_f = 0$ and it follows that

$$\begin{aligned}
\tfrac{\partial}{\partial t} g((\Phi_t)_* X_0, (\Phi_t)_* Y_0) &= \tfrac{\partial}{\partial t} g(X \circ \Phi_t, Y \circ \Phi_t) \\
&= ((\nabla f) g(X, Y)) \circ \Phi_t \\
&= (g(\nabla_{\nabla f} X, Y) + g(X, \nabla_{\nabla f} Y)) \circ \Phi_t \\
&= (g(h_f(X), Y) + g(X, h_f(Y))) \circ \Phi_t \\
&= 2H_f(X, Y) \circ \Phi_t = 0.
\end{aligned}$$

Thus $g((\Phi_t)_* X_0, (\Phi_t)_* Y_0)$ is constant for all $t \in \mathbf{R}$ and it follows that $\Phi_t \mid_{f^{-1}(0)}$ is an isometry for each $t \in \mathbf{R}$. Hence $g_t = g_0$ for all t and (M, g) is isometric to (M_k, g_k) $= (\mathbf{R} \times N, kdt \otimes dt \oplus g_N)$, where $g_N = g_0$. □

Now we consider the more general case of f satisfying either timelike or spacelike eikonal inequality rather than timelike or spacelike eikonal equations.

Theorem 6.3.2 (The Generalized Semi-Riemannian Regular Interval Theorem) *Let $f : (M, g) \to \mathbf{R}$ be a map, where M is connected. If f satisfies the timelike or spacelike eikonal inequality on (M, g) (that is, if $g(\nabla f, \nabla f) < 0$ or $g(\nabla f, \nabla f) > 0$) and $\overset{c}{\nabla} f$ is a complete vector field on M, where $\overset{c}{\nabla}$ is the gradient with respect to conformal metric tensor $g_c = | g(\nabla f, \nabla f) | g$, then (M, g) is conformally diffeomorphic to a parametrized semi-Riemannian product manifold (M_k, g_k) $= (\mathbf{R} \times N, kdt \otimes dt \oplus g_{ct})$, where $k = -1$ if $g(\nabla f, \nabla f) < 0$ and $k = 1$ if $g(\nabla f, \nabla f)$ > 0. Moreover, if f is affine on (M, g_c) as well, then (M, g) is conformally diffeomorphic to semi-Riemannian product manifold $(M_k, g_k) = (\mathbf{R} \times N, kdt \otimes dt \oplus g_N)$, where $k = -1$ if $g(\nabla f, \nabla f) < 0$ and $k = 1$ if $g(\nabla f, \nabla f) > 0$.*

Proof. First note that, by Proposition 6.1.4, f satisfies a nonnull eikonal equation $g_c(\overset{c}{\nabla} f, \overset{c}{\nabla} f) = k$ on (M, g_c), where $k = -1$ if $g(\nabla f, \nabla f) < 0$ and $k = 1$ if $g(\nabla f, \nabla f)$ > 0. Also since $\overset{c}{\nabla} f$ is complete, it follows from Theorem 6.3.1 that (M, g_c) is isometric to $(M_k, g_k) = (\mathbf{R} \times N, kdt \otimes dt \oplus g_{ct})$. In particular, if f is affine on (M, g_c) as well, (M, g_c) is isometric to $(M_k, g_k) = (\mathbf{R} \times N, kdt \otimes dt \oplus g_N)$. Since $g_c =$ $| g(\nabla f, \nabla f) | g$, this isometry is then a conformal diffeomorphism from (M, g) onto (M_k, g_k). That is, (M, g) and (M_k, g_k) are conformally diffeomorphic. □

In the statement of Theorem 6.3.1, if we assume that (M, g) is geodesically complete then there is no need to assume that ∇f is a complete vector field on M. Indeed, since ∇f is a geodesic vector field on (M, g) by Proposition 6.1.2, ∇f is complete by the geodesic completeness of (M, g). But in the statement of Theorem 6.3.2, even if we assume that (M, g) is geodesically complete, (M, g_c) may not be geodesically complete and hence $\overset{c}{\nabla} f$ may not be complete although it is a geodesic vector field on (M, g_c) by Proposition 6.1.2. But in the Riemannian case, the assumption that $g(\nabla f, \nabla f)$ bounded away from zero reflects the geodesic completeness of (M, g) to (M, g_c). Next we will state the generalized Riemannian regular interval theorem from this point of view.

Lemma 6.3.2 [F2] *Let (M, g) be a complete, connected Riemannian manifold. Let $g_c = \varphi g$ be a conformal metric tensor to g with $\varphi \geq r^2 > 0$ for some $r \in (0, \infty)$. Then (M, g_c) is also a complete Riemannian manifold.*

Proof. Let $\gamma : [t_1, t_2] \to M$ be a curve from p to q and, let $L(\gamma)$ and $L_c(\gamma)$ denote its lengths in (M, g) and (M, g_c), respectively. Then

$$\begin{aligned} L_c(\gamma) &= \int_{t_1}^{t_2} g_c(\dot{\gamma}, \dot{\gamma})^{1/2} dt \\ &= \int_{t_1}^{t_2} \varphi(\gamma(t))^{1/2} g(\dot{\gamma}, \dot{\gamma})^{1/2} dt \\ &\geq r \int_{t_1}^{t_2} g(\dot{\gamma}, \dot{\gamma})^{1/2} dt \\ &\geq rL(\gamma). \end{aligned}$$

Applying this inequality to each piece of a piecewise smooth curve shows that the same inequality remains true for piecewise smooth curves.

Let d and d_c denote the Riemannian distance functions of (M, g) and (M, g_c), respectively. Then

$$d_c(p, q) = \inf_{\{\gamma\}} L_c(\gamma) \geq r \inf_{\{\gamma\}} L(\gamma) = rd(p, q),$$

where the infimun is taken over all piecewise smooth curves $\{\gamma\}$ from p to q.

Now let $\{p_n\}$ be a Cauchy sequence in (M, d_c). Then for any $\varepsilon > 0$, there exists $N > 0$ such that for $m, n > N$,

$$rd(p_m, p_n) \leq d_c(p_m, p_n) < r\varepsilon.$$

Thus $d(p_m, p_n) < \varepsilon$ so $\{p_n\}$ is a Cauchy sequence in (M, d). Since (M, g) is a complete Riemannian manifold, $\{p_n\}$ converges in M. Thus by Hopf-Rinow theorem (see Theorem B.1.1), (M, g_c) is a complete Riemannian manifold. □

Now by using this observation, we state the Generalized Riemannian Regular Interval Theorem.

Theorem 6.3.3 [F2] (The Generalized Riemannian Regular Interval Theorem) *Let (M, g) be a complete, connected Riemannian manifold and $f : (M, g) \to \mathbf{R}$ be a map with $g(\nabla f, \nabla f) \geq r^2 > 0$, where $r \in (0, \infty)$. Then (M, g) is conformally diffeomorphic to a parametrized Riemannian product manifold $(M_1, g_1) = (\mathbf{R} \times N, dt \otimes dt \oplus g_{ct})$, where $g_c = g(\nabla f, \nabla f)g$. Moreover, if f is affine on (M, g_c) as well, where $g_c = g(\nabla f, \nabla f)g$, then (M, g) is conformally diffeomorphic to a Riemannian product manifold $(M_1, g_1) = (\mathbf{R} \times N, dt \otimes dt \oplus g_N)$.*

Proof. Let $g_c = g(\nabla f, \nabla f)g$ be a conformal metric tensor to g. Then by Lemma 6.3.2, (M, g_c) is also a complete Riemannian manifold. Hence since $\overset{c}{\nabla} f$ is a geodesic vector field on (M, g_c) by Proposition 6.1.2, the geodesic completeness of (M, g_c) implies that $\overset{c}{\nabla} f$ is a complete vector field on M, where $\overset{c}{\nabla}$ is the gradient with respect to a conformal metric tensor $g_c = g(\nabla f, \nabla f)g$. Thus the conclusion of the theorem follows from Theorem 6.3.2. □

Chapter 7

Applications To Splitting Theorems

As we know from the semi-Riemannian regular interval theorem (Theorem 6.3.1), the existence of an affine solution f of a nonnull eikonal equation on a semi-Riemannian manifold (M, g) yields a splitting of (M, g) into a semi-Riemannian product manifold, provided that ∇f is a complete vector field on (M, g). We also know from Proposition 6.2.5 that being affine for a solution f of a semi-Riemannian eikonal equation is related to the Ricci curvature of (M, g) and the Hessian tensor of f. Now, by combining these results, one can obtain splitting theorems for certain semi-Riemannian manifolds which admit solutions to nonnull eikonal equations. In fact, in this chapter we obtain such splitting theorems in a more general context for semi-Riemannian, Riemannian and Lorentzian manifolds. We devote sections 7.1, 7.2 and 7.3 to such splitting theorems for semi-Riemannian, Riemannian and Lorentzian manifolds, respectively. Section 7.3 may be considered of separate interest for applications of semi-Riemannian maps in General Relativity.

The reader who is not familiar to the global concepts in Riemannian and/or Lorentzian geometry is referred to Appendix B.

7.1 The Semi-Riemannian Splitting Theorem

Throughout this section, let (M, g) be a semi-Riemannian manifold of index ν and dimension n.

Definition 7.1.1 Let $f : (M, g) \to \mathbf{R}$ be a map. The *energy density* of f is defined by

$$e(f) = -\frac{1}{2} g(\nabla f, \nabla f).$$

Also f is called of *subharmonic energy density* if $e(f)$ is a subharmonic map on (M, g), that is, $-\Delta e(f) \geq 0$.

Remark 7.1.1 It is worth emphasizing that, if $f : (M, g) \to \mathbf{R}$ satisfies a semi-Riemannian eikonal equation on (M, g) then f is of subharmonic energy density, that is, $-\Delta e(f) = 0$.

Theorem 7.1.1 (The Semi-Riemannian Splitting Theorem). *Let (M, g) be a connected semi-Riemannian manifold of dimension ≥ 2 which satisfies the following conditions:*

(1) *(M, g) admits a map $f : (M, g) \to \mathbf{R}$ with $g(\nabla f(p), \nabla f(p)) \neq 0$ at some $p \in M$, which is of subharmonic energy density and has diagonalizable Hessian tensor at each $p \in M$,*

(2) *$Ric(\nabla f, \nabla f) \geq 0$,*

(3) *∇f is a complete vector field on M.*

Then (M, g) is isometric to a semi-Riemannian product manifold $(M_k, g_k) = (\mathbf{R} \times N, kdt \otimes dt \oplus g_N)$, where $k = -1$ if $g(\nabla f(p), \nabla f(p)) < 0$ and $k = 1$ if $g(\nabla f(p), \nabla f(p)) > 0$.

Proof. Since $\Delta e(f) \leq 0$, by Corollary 3.4.1,

$$0 \geq Ric(\nabla f, \nabla f) + \|h_f\|^2 - g(\nabla \Delta f, \nabla f).$$

Thus, it follows from $Ric(\nabla f, \nabla f) \geq 0$ and h_f is diagonalizable at each $p \in M$ (see Lemma 1.2.1) that

$$g(\nabla \Delta f, \nabla f) \geq \|h_f\|^2 \geq \frac{1}{n}(\operatorname{trace} h_f)^2 = \frac{1}{n}(\Delta f)^2.$$

Now by using completeness of ∇f, it can be shown as in the proof of Proposition 6.2.5 that $\Delta f = 0$ on (M, g). Hence $\|h_f\|^2 = 0$ and since h_f is diagonalizable at each $p \in M$, $h_f = 0$. Thus f is an affine map on (M, g). But then, $0 \neq g(\nabla f, \nabla f)$ is constant on M by Proposition 6.2.2–(2), and the map $h : (M, g) \to \mathbf{R}$ defined by

$$h = \frac{1}{|g(\nabla f, \nabla f)|^{1/2}} f$$

is an affine solution of a nonnull eikonal equation on (M, g). (See Proposition 6.2.3). Also since h is a scalar multiple of f, ∇h is complete on M. Hence by Theorem 6.3.1, (M, g) is isometric to a semi-Riemannian product manifold $(M_k, g_k) = (\mathbf{R} \times N, dt \otimes dt \oplus g_N)$, where $k = -1$ if $g(\nabla h, \nabla h) = -1$ and $k = 1$ if $g(\nabla h, \nabla h) = 1$. □

7.2 Riemannian Splitting Theorems

In Riemannian geometry, there is a well-known splitting theorem due to Cheeger and Gromoll. If we call an inextendable minimal geodesic of a Riemannian manifold a *line* (see Remark B.1.1), then the Riemannian splitting theorem can be stated as follows:

Theorem 7.2.1 [CG] **(The Riemannian Splitting Theorem).** *Let (M, g) be a complete, connected Riemannian manifold of dimension ≥ 2 which satisfies the curvature condition*

$$Ric(v, v) \geq 0 \quad \text{for all} \quad v \in TM$$

and which contains a line. Then (M, g) is isometric to a Riemannian product manifold $(M_1, g_1) = (\mathbf{R} \times N, dt \otimes dt \oplus g_N)$.

Now we state the Riemannian version of Theorem 7.1.1 as another Riemannian splitting theorem.

Theorem 7.2.2 *Let (M, g) be a connected Riemannian manifold of dimension ≥ 2 which satisfies the following conditions:*

(1) *(M, g) admits a map $f : (M, g) \to \mathbf{R}$ with $g(\nabla f(p), \nabla f(p)) > 0$ at some $p \in M$, which is of subharmonic energy density,*

(2) *$Ric(\nabla f, \nabla f) \geq 0$,*

(3) *∇f is a complete vector field on M.*

Then (M, g) is isometric to a Riemannian product manifold $(M_1, g_1) = (\mathbf{R} \times N, dt \otimes dt \oplus g_N)$.

Proof. Immediate from Theorem 7.1.1 since h_f is diagonalizable at each $p \in M$ by the assumption that (M, g) is a Riemannian manifold. □

Remark 7.2.1 Note that, in Theorem 7.2.1, (N, g_N) is a complete Riemannian manifold. But in Theorem 7.2.2, (N, g_N) is not necessarily a complete Riemannian manifold.

Now we state a variation of Theorem 7.2.2 for complete Riemannian manifolds. For this, we need a fact about completeness of vector fields on a complete Riemannian manifold whose norms are bounded from above. For this, first we make the following observation.

Lemma 7.2.1 *Let (M, g) be a complete Riemannian manifold and let $\gamma : [a, b) \to M$ be a curve of finite length in (M, g). Then there exist a point $p \in M$ such that $\gamma(t) \to p$ as $t \to b^-$.*

Proof. Let d be the Riemannian distance function of (M, g) and let $L(\gamma)$ be the length of γ in (M, g). If we set $K = \{q \in M/d(\gamma(0), q) \leq L(\gamma)\}$, then by Hopf-Rinow theorem (see Theorem B.1.1), K is compact. Let $\{t_n\}$ be a sequence in $[a, b)$ with $t_n \to b$. Since $d(\gamma(0), \gamma(t)) \leq L(\gamma \mid [a, t)) \leq L(\gamma)$ for $t \in [a, b)$, we have $\gamma([a, b)) \subseteq K$. Thus by compactness of K, the sequence $\{\gamma(t_n)\}$ has a limit point $p \in K$. If $\lim_{t \to b^-} \gamma(t) \neq p$, there would then exist an $\varepsilon > 0$ such that γ leaves the ball $B(m, \varepsilon) = \{m \in M/d(p, m) \leq \varepsilon\}$ infinitely often. But this would imply that γ has infinite length, in contradiction. Thus $\lim\limits_{t \to b^-} \gamma(t) = p$. □

Proposition 7.2.1 *Let (M, g) be a complete Riemannian manifold. If Z is a vector field on M with $g(Z, Z) \leq r^2$, where $r \in [0, \infty)$, then Z is complete.*

Proof. Let $\gamma : [a, b) \to M$ be an integral curve of Z which is inextendable beyond $b < \infty$. (The case $\gamma : (a, b] \to M$ is inextendable beyond $a > -\infty$ is similar). Then since $g(\dot{\gamma}, \dot{\gamma})^{1/2} = g(Z \circ \gamma, Z \circ \gamma)^{1/2} \leq r$, γ has finite length $L(\gamma) \leq r(b - a)$ in (M, g) and hence, $\lim\limits_{t \to b^-} \gamma(t) = p \in M$ by Lemma 7.2.1. But this is contradiction with the fact that γ is inextendable beyond b. Thus $b = \infty$. □

Theorem 7.2.3 *Let (M, g) be a complete, connected Riemannian manifold of dimension ≥ 2 which satisfies the following conditions:*

(1) *(M, g) admits a map $f : (M, g) \to \mathbf{R}$ with $g(\nabla f(p), \nabla f(p)) > 0$ at some $p \in M$ and $g(\nabla f, \nabla f) \leq r^2$, where $r \in (0, \infty)$, which is of subharmonic energy density,*

(2) *$Ric(\nabla f, \nabla f) \geq 0$.*

Then (M, g) is isometric to a Riemannian product manifold $(M_1, g_1) = (\mathbf{R} \times N, dt \otimes dt \oplus g_N)$.

Proof. By Proposition 7.2.1, since ∇f is a complete vector field on M, the conclusion of the theorem immediately follows from Theorem 7.2.2. □

Next we state another Riemannian splitting theorem involving curvatures and Riemannian maps. Recall that if (M_1, g_1) and (M_2, g_2) are Riemannian manifolds then the *Riemannian product* of (M_1, g_1) and (M_2, g_2) is defined to be the product manifold $M = M_1 \times M_2$ with metric tensor $g = g_1 \oplus g_2$. Here, for x, $y \in T_{(p_1, p_2)}M$,

$$g(x, y) = g_1(\pi_{1*}x, \pi_{1*}y) + g_2(\pi_{2*}x, \pi_{2*}y),$$

where π_1 and π_2 are the projections of M onto M_1 and M_2, respectively.

To obtain the mentioned Riemannian splitting theorem, we need a Corollary of Vilms about affine Riemannian submersions. We do not provide the proof of this corollary here since it involves certain concepts we have not introduced in this book so as not to disturb the elementary level of the preliminaries. Interested readers are invited to refer to [V] for its proof.

Theorem 7.2.4 [V] *Let (M_1, g_1) and (M_2, g_2) be Riemannian manifolds with* $\dim M_1 = n_1 > \dim M_2 = n_2 \geq 1$, *where (M_1, g_1) is complete, connected and simply connected. If there is a surjective affine Riemannian submersion $f : (M_1, g_1) \to (M_2, g_2)$ then (M_1, g_1) is isometric to a Riemannian product manifold $(M, g) = (M_2 \times M_3, g_2 \oplus g_3)$, where (M_3, g_3) is a Riemannian manifold.*

Now by using the notation of Section 4.4, we state the mentioned splitting theorem as follows:

Theorem 7.2.5 [GK2] *Let (M_1, g_1) be a complete, connected and simply connected Riemannian manifold with $r_1 \geq A$ and let (M_2, g_2) be a Riemannian manifold with $S_2 \leq B$ such that $n_1 > n_2 \geq 1$ and $n_2 A \geq B$. If there is a surjective Riemannian map $f : (M_1, g_1) \to (M_2, g_2)$ with* $\operatorname{div} \tau(f) \geq 0$ *then (M_1, g_1) is isometric to a Riemannian product manifold $(M, g) = (M_2 \times M_3, g_2 \oplus g_3)$, where (M_3, g_3) is a Riemannian manifold.*

Proof. First note that since f is a surjective Riemannian map, by Theorem 4.3.1-(2), f is a surjective Riemannian submersion. Also by Theorem 4.4.4, f is affine and hence f is a surjective affine Riemannian submersion. Hence the conclusion of the theorem follows from Theorem 7.2.4. □

7.3 Lorentzian Splitting Theorems

After a sequence of papers by Beem, Ehrlich, Markvorsen and Galloway [BEMG], Eschenburg [E], Galloway [G] and Newman [Ne], the Lorentzian splitting theorem was obtained as an analog of the Riemannian splitting theorem (cf. [BEE, Chapter 14]). If we call an inextendable maximal timelike geodesic on a Lorentzian manifold a *timelike line* (see Remark B.2.3), then the Lorentzian splitting theorem can be stated as follows:

Theorem 7.3.1 [BEE] **(The Lorentzian Splitting Theorem).** *Let (M, g) be a spacetime of dimension $n > 2$ which satisfies the following conditions:*

(1) *(M, g) is either timelike geodesically complete or globally hyperbolic,*

(2) *$Ric(v, v) \geq 0$ for every timelike $v \in TM$,*

(3) *(M, g) contains a complete timelike line.*

Then (M, g) is isometric to a Lorentzian product manifold $(M_{-1}, g_{-1}) = (\mathbf{R} \times N, -dt \otimes dt \oplus g_N)$, where (N, g_N) is a complete Riemannian manifold.

Here we first state the Lorentzian version of Theorem 7.1.1 for stably causal spacetimes as another Lorentzian splitting theorem. Recall that a Lorentzian manifold (M, g) is *stably causal* if and only if (M, g) admits a map $f : (M, g) \to \mathbf{R}$ satisfying the timelike eikonal inequality $g(\nabla f, \nabla f) < 0$. (See Proposition B.2.4).

Theorem 7.3.2 *Let (M, g) be a connected Lorentzian manifold of dimension $n \geq 2$ which satisfies the following conditions:*

(1) *(M, g) admits a map $f : (M, g) \to \mathbf{R}$ satisfying the timelike eikonal inequality $g(\nabla f, \nabla f) < 0$, which is of subharmonic energy density and has diagonalizable Hessian tensor at each $p \in M$,*

(2) *$Ric(\nabla f, \nabla f) \geq 0$,*

(3) *∇f is a complete vector field on M.*

Then (M, g) is isometric to a Lorentzian product manifold $(M_{-1}, g_{-1}) = (\mathbf{R} \times N, -dt \otimes dt \oplus g_N)$, where (N, g_N) is a Riemannian manifold.

Proof. Immediate from Theorem 7.1.1. □

Remark 7.3.1 Note that, in Theorem 7.3.1, (N, g_N) is a complete Riemannian manifold. But in Theorem 7.3.2, (N, g_N) is not necessarily a complete Riemannian manifold.

The above theorem can also be stated similar to a rigid singularity theorem for stably causal spacetimes as follows:

Corollary 7.3.1 *Let (M, g) be a connected Lorentzian manifold of dimension $n \geq 2$ which satisfies the following conditions:*

(1) *(M, g) admits a map $f : (M, g) \to \mathbf{R}$ satisfying the timelike eikonal inequality $g(\nabla f, \nabla f) < 0$, which is of subharmonic energy density and has diagonalizable Hessian tensor at each $p \in M$,*

(2) $Ric(\nabla f, \nabla f) \geq 0$.

Then either ∇f is incomplete on M, or else (M,g) is isometric to a Lorentzian product manifold $(M_{-1}, g_{-1}) = (\mathbf{R} \times N, -dt \otimes dt \oplus g_N)$, where (N, g_N) is a Riemannian manifold.

A nontrivial example to the above corollary is the Schwarzchild black hole. Using the notation of O'Neill [O, Chapter 13], for the Schwarzchild radius function

$$r : \mathbf{R} \times (0, 2\mathrm{M}) \times S^2 \to (0, 2\mathrm{M}),$$

we have

$$g(\nabla r, \nabla r) = (1 - \frac{2\mathrm{M}}{r}) < 0$$

and

$$h_r(X) = \begin{cases} \frac{\mathrm{M}}{r^2} X & \text{for every vector field } X \text{ orthogonal to } S^2(r) \\ \frac{1}{r}(1 - \frac{2\mathrm{M}}{r})X & \text{for every vector field } X \text{ tangent to } S^2(r) \end{cases},$$

where $S^2(r) = \pi^{-1}(t,r)$ and $\pi : \mathbf{R} \times (0, 2\mathrm{M}) \times S^2 \to \mathbf{R} \times (0, 2\mathrm{M})$ is the projection. (That is, h_r is diagonalizable at each point). Also

$$-\Delta e(r) = \frac{2\mathrm{M}^2}{r^4} > 0.$$

Hence the above corollary implies that ∇r is incomplete since the Schwarzchild black hole is not isometric to a Lorentzian product $(M_{-1}, g_{-1}) = (\mathbf{R} \times N, -dt \otimes dt \oplus g_N)$. On the other hand, note that the Schwarzchild time function

$$t : \mathbf{R} \times (2\mathrm{M}, \infty) \times S^2 \to \mathbf{R}$$

on the exterior Schwarzchild spacetime has nondiagonalizable Hessian tensor h_t at each point.

Note that a Lorentzian manifold (M,g) is stably causal if and only if timelike eikonal inequality has a solution on (M,g). But in general, on a stably causal Lorentzian manifold (M,g), timelike eikonal equation may not have a solution, that is, there may not exist a semi-Riemannian submersion $f : (M,g) \to (\mathbf{R}, -dt \otimes dt)$ with fibers as semi-Riemannian submanifolds (in fact, spacelike hypersurfaces) in

(M, g). (See Proposition 6.1.1–(1)). But the existence of a solution of timelike eikonal equation on a Lorentzian manifold has an important consequence. Note that, if $f : (M, g) \to \mathbf{R}$ is a solution of timelike eikonal equation $g(\nabla f, \nabla f) = -1$ on a Lorentzian manifold (M, g), then ∇f is a unit timelike geodesic vector field on (M, g). (See Proposition 6.1.2). Hence $h_f(\nabla f) = 0$, that is, ∇f is a unit eigenvector field of h_f corresponding to zero eigenfunction. Therefore the orthogonal bundle ker f_* to span $\{\nabla f\}$ is an invariant subbundle of h_f with induced positive definite metric tensor. Thus the restriction of h_f to ker f_* is also diagonalizable at each point and hence, h_f is diagonalizable at each $p \in M$. There is also another important property of the solutions of timelike eikonal equation on a Lorentzian manifold. Recall that a connected, time oriented Lorentzian manifold is called a spacetime. (See Definition B.2.1).

Definition 7.3.1 Let S be an acausal set in a spacetime (M, g). A future-directed causal curve $\gamma : I \to M$, where I is an interval, with $\gamma(t_0) \in S$ for $t_0 \in I$ is said to *realize the distance between S and its points* if, for every $t_0 < t \in I$, $L(\gamma \mid [t_0, t]) = d(S, \gamma(t))$, and for every $t_0 > t \in I$, $L(\gamma \mid [t, t_0]) = d(\gamma(t), S)$, where L denotes the length in (M, g) and d is the Lorentzian distance function of (M, g).

Remark 7.3.2 Note that if $\gamma : I \to M$ is a future-directed causal curve realizing the distance between S and its points then necessarily γ is a geodesic. (See Appendix B). But γ may fail to be a maximal geodesic.

Lemma 7.3.1 *Let (M, g) be a spacetime and $f : (M, g) \to \mathbf{R}$ be a map satisfying the timelike eikonal equation on (M, g) with future-directed ∇f. Then every integral curve of ∇f realizes the distance between each level set of f it meets and its points, and hence, in particular, every integral curve of ∇f is a maximal geodesic.*

Proof. First note that, since $(f \circ \alpha)' = g((\nabla f) \circ \alpha, \dot{\alpha}) < 0$ for any future-directed causal curve α, f is decreasing along future-directed causal curves. Hence in particular, level sets of f are acausal. Now let $\gamma : I \to M$ be an integral curve of ∇f passing through $p \in f^{-1}(t_0) \neq \emptyset$ with $\gamma(t_0) = p$. We will show that $d(f^{-1}(t_0), \gamma(t))$ $= L(\gamma \mid (t_0, t])$ for all $t_0 < t \in I$. (The case $d(\gamma(t), f^{-1}(t_0)) = L(\gamma \mid (t, t_0])$ for all $t_0 > t \in I$ is similar). Let $\alpha : [s_1, s_2] \to M$ be a future-directed causal curve with $\alpha(s_1) \in f^{-1}(t_0)$ and $\alpha(s_2) = \gamma(t)$ for some $t_0 < t \in I$. Then by the wrong-way Cauchy-Schwarz inequality (see [SW, p. 24])

$$\begin{aligned} |g((\nabla f) \circ \alpha, \dot{\alpha})| &\geq |g(\dot{\alpha}, \dot{\alpha})|^{1/2} |g((\nabla f) \circ \alpha, (\nabla f) \circ \alpha)|^{1/2} \\ &= |g(\dot{\alpha}, \dot{\alpha})|^{1/2}, \end{aligned}$$

we have

$$\begin{aligned} L(\alpha) &= \int_{s_1}^{s_2} |g(\dot{\alpha},\dot{\alpha})|^{1/2}\, ds \\ &\leq \int_{s_1}^{s_2} |g(((\nabla f)\circ\alpha),\dot{\alpha})|\, ds \\ &= -\int_{s_1}^{s_2} (f\circ\alpha)' ds \\ &= -f(\alpha(s_2)) + f(\alpha(s_1)) \\ &= -f(\gamma(t)) + t_0 \\ &= t - t_0 = L(\gamma \mid [t_0, t]). \end{aligned}$$

Thus γ realizes the distance between $f^{-1}(t_0)$ and its points for any $t_0 \in$ image (f). Hence, in particular, γ is a maximal geodesic. □

A causal spacetime (M, g) is called *finitely compact* if for each fixed constant $B > 0$ and each sequence of points $\{x_n\}$ with either $p \ll q \leq x_n$ and $d(p, x_n) \leq B$ for all n, or $x_n \leq q \ll p$ and $d(x_n, p) \leq B$ for all n, there is a point of accumulation of $\{x_n\}$ in M, where d is the Lorentzian distance function of (M, g). Also a causal spacetime (M, g) is called *timelike Cauchy complete* if any sequence $\{x_n\}$ of points with $x_n \ll x_{n+m}$ for $n, m = 1, 2, \ldots$ and $d(x_n, x_{n+m}) \leq B_n$ (or else, $x_{n+m} \ll x_n$ for $n, m = 1, 2, \ldots$ and $d(x_{n+m}, x_n) \leq B_n$) for all $m \geq 1$, where $B_n \to 0$ as $n \to \infty$, is a convergent sequence. Finally we define a condition on a causal spacetime (M, g) as follows: Let $p \ll q$ (or $q \ll p$) and let $\gamma : [0, b) \to M$ be a future (past) inextendable future-directed (past-directed) causal geodesic starting at q with parameter t defined for $0 < t < b \leq \infty$. Then (M, g) is said to satisfy *Condition A* if $d(p, \gamma(t)) \to \infty$ $(d(\gamma(t), p) \to \infty)$ as $t \to b$ for all such p, q and $\gamma(t)$. (See [B], [BEE, p. 211]). In fact, finite compactness, timelike Cauchy completeness and Condition A in Lorentzian geometry are the analogs of finite compactness, Cauchy completeness and geodesic completeness in Riemannian geometry, respectively. Although these three conditions in Lorentzian geometry are equivalent in globally hyperbolic spacetimes, they do not imply timelike geodesic completeness even for the class of globally hyperbolic spacetimes. (But note that these three conditions in Lorentzian geometry are not equivalent in general). However, as we see in the lemma below, each of these conditions in Lorentzian geometry has a relation to timelike geodesic completeness.

Lemma 7.3.2 *Let (M, g) be a strongly causal finitely compact or a strongly causal timelike Cauchy complete or a causal spacetime satisfying Condition A. If $\gamma : (a, b) \to M$ is an inextendable maximal future-directed timelike geodesic, (that is, γ is a timelike line) then γ is complete.*

Proof. First suppose that (M, g) is finitely compact and assume $b < \infty$. (The case, $a > -\infty$ is similar). Then for given $B > 0$, there exists $t_0 \in (a, b)$ such that $L(\gamma \mid [t_0, b)) < B$. Now let $\{t_n\}$ be a sequence of points in (a, b) with $t_0 < t_0' \leq t_n$ and $t_n \to b$. Then, if we set $p = \gamma(t_0)$, $q = \gamma(t_0')$ and $x_n = \gamma(t_n)$, we have $p \ll q \leq x_n$ and $d(p, x_n) = L(\gamma \mid [t_0, t_n]) \leq L(\gamma \mid [t_0, b)) < B$. Thus, the finite compactness of (M, g) implies that $\{x_n\}$ has an accumulation point $x \in M$. But since (M, g) is strongly causal, by the existence of a causally convex neighborhood base at x, $\lim_{t \to b^-} \gamma(t) = x$ in contradiction to the future inextendability of γ. Thus $b = \infty$.

Now suppose that (M, g) is timelike Cauchy complete and assume $b < \infty$. (The case, $a > -\infty$ is similar). Let $B_n = \frac{1}{n}$ and let $t_1 < t_2 < \cdots < t_n < \cdots$ be points in (a, b) with $t_n \to b$ and $L(\gamma \mid [t_n, b)) < \frac{1}{n}$. Then, if we set $x_n = \gamma(t_n)$, we have $x_n \ll x_{n+m}$ for $n, m = 1, 2, \ldots$ and $d(x_n, x_{n+m}) = L(\gamma \mid [t_n, t_{n+m}]) < L(\gamma \mid [t_n, b)) < \frac{1}{n}$ $= B_n$ for all $m \geq 1$. Thus the timelike Cauchy completeness of (M, g) implies that $\{x_n\}$ has an accumulation point $x \in M$. But since (M, g) is strongly causal, by the existence of a causally convex neighborhood base at x, $\lim_{t \to b^-} \gamma(t) = x$ in contradiction to the future inextendability of γ. Thus $b = \infty$.

Finally suppose that (M, g) satisfies Condition A and assume $b < \infty$. (The case, $a > -\infty$ is similar). Without loss of generality we may assume that $0 \in (a, b)$ and let $0 > t_0 \in (a, b)$. Now set $p = \gamma(t_0)$ and $q = \gamma(0)$. Then $p \ll q$ and $\gamma \mid [0, b)$ is a future inextendable future-directed timelike geodesic starting at q such that $d(p, \gamma(t)) = L(\gamma \mid [t_0, t]) < L(\gamma \mid [t_0, b)) < \infty$ for all $t \in (0, b)$. But since (M, g) satisfies Condition A, this gives a contradiction as $t \to b$. Thus $b = \infty$. □

Now by using Lemma 7.3.1 and 7.3.2, we state the following version of Theorem 7.3.2 for either a finitely compact spacetime or a timelike Cauchy complete spacetime or a spacetime satisfying Condition A. Recall that a stably causal spacetime is strongly causal (see Proposition B.2.7) and a strongly causal spacetime is causal. (See Proposition B.2.6).

Theorem 7.3.3 [GK3] *Let (M, g) be a spacetime of dimension ≥ 2 which satisfies the following conditions:*

(1) *(M, g) admits a map $f : (M, g) \to \mathbf{R}$ satisfying the timelike eikonal equation,*

(2) *$Ric(\nabla f, \nabla f) \geq 0$,*

(3) *(M, g) is either finitely compact or timelike Cauchy complete or satisfying Condition A.*

Then (M, g) is isometric to a Lorentzian product $(M_{-1}, g_{-1}) = (\mathbf{R} \times N, -dt \otimes dt \oplus g_N)$, where (N, g_N) is a complete Riemannian manifold.

Proof. First note that (M, g) is stably causal by (1) and without loss of generality, we may assume that ∇f is future-directed. Also recall that, since ∇f is a geodesic vector field by Proposition 6.1.2 and $\ker f_*$ has induced positive definite metric tensor, h_f is diagonalizable at each $p \in M$. Furthermore, since $e(f) = \frac{1}{2}$, f is of subharmonic energy density. Also since the inextendable integral curves of ∇f are timelike lines by Lemma 7.3.1, it follows from Lemma 7.3.2 that they are complete, and hence ∇f is a complete vector field. Thus it follows from Theorem 7.3.2 that (M, g) is isometric to a Lorentzian product manifold $(M_{-1}, g_{-1}) = (\mathbf{R} \times N, -dt \otimes dt \oplus g_N)$, where (N, g_N) is a Riemannian manifold. Now it remains to show that either of finite compactness or timelike Cauchy completeness or Condition A imply that (N, g_N) is complete.

First suppose that (M, g) is finitely compact. Now to show that (N, g_N) is complete, it suffices to show that every open metric ball $B_N(y, \epsilon) = \{y' \in N / d_N(y, y') < \epsilon\}$ in (N, g_N) has compact closure, where d_N is the Riemannian distance function of (N, g_N). Then it follows from the Hopf-Rinow theorem (see Theorem B.1.1) that (N, g_N) is complete. Now let $\{y_n\}$ be a sequence of points in $B_N(y, \epsilon)$. Let $\epsilon_1 > 0$ and put $B = \epsilon + \epsilon_1$. Now set $p = (0, y)$, $q = (\epsilon_1, y)$ and $x_n = (B, y_n)$ in $\mathbf{R} \times N$. First observe that $q \ll x_n$ for all n. To see this, let $\alpha_n : [0, s_n] \to B_N(y, \epsilon)$ be a unit speed curve in (N, g_N) with $\alpha_n(0) = y$, $\alpha_n(s_n) = y_n$ and $s_n < \epsilon$. Then it is easy to see that the curve $\beta_n : [0, s_n] \to \mathbf{R} \times N$ given by $\beta_n(s) = (\frac{\epsilon}{s_n} s + \epsilon_1, \alpha_n(s))$ is a future-directed timelike curve with $\beta_n(0) = q$ and $\beta_n(s_n) = x_n$ for all n. Also $p \ll q$ and $d(p, x_n) \leq B$ for all n since by Lemma 7.3.1, the integral curve of ∇f passing through p realizes the distance between the level set $\{B\} \times N$ and its points. Thus $p \ll q \leq x_n$ and $d(p, x_n) \leq B$ for all n. Then the finite compactness of $(M_{-1}, g_{-1}) = (\mathbf{R} \times N, -dt \otimes dt \oplus g_N)$ implies that $\{x_n\}$ has an accumulation point in $\mathbf{R} \times N$ and hence, $\{y_n\}$ has an accumulation point in the closure of $B_N(y, \epsilon)$. Thus the closure of $B_N(y, \epsilon)$ is compact for all $y \in N$ and $\epsilon > 0$.

Now suppose that (M, g) is timelike Cauchy complete. To show that (N, g_N) is complete, it suffices to show that every Cauchy sequence in (N, d_N) has a convergent subsequence. Then it follows from the Hopf-Rinow theorem that (N, g_N) is complete. Let $\{z_k\}$ be a Cauchy sequence in (N, d_N). Then for $C_n = \frac{1}{n(n+1)}$, let $\{z_{k_n} = y_n\}$ be a subsequence of $\{z_k\}$ such that $d_N(y_n, y_{n+m}) < C_n$ for all $m \geq 1$. (We can choose this subsequence $\{y_n\}$ as follows: Since $\{z_k\}$ is a Cauchy sequence, for each $C_n = \frac{1}{n(n+1)}$, there exists $K_n > 0$ such that $d_N(z_k, z_l) < C_n$ for every $k, l \geq K_n$. Put $z_{K_n} = y_n$ inductively such that $K_1 < K_2 < \cdots < K_n < \cdots$. Then clearly the subsequence $\{y_n\}$ of $\{z_k\}$ satisfies $d(y_n, y_{n+m}) < C_n$ for all $m \geq 0$). Now put $x_n = (B_n, y_n)$ in $\mathbf{R} \times N$, where $B_n = \frac{1}{n}$. First note that $x_{n+m} \ll x_n$ for $n, m = 1, 2, \ldots$. To see this, let $\alpha_{n,m} : [0, s_{n,m}] \to N$ be a unit speed curve in (N, g_N) with $\alpha_{n,m}(0) = y_n$, $\alpha_{n,m}(s_{n,m}) = y_{n+m}$ and $s_{n,m} < C_n$. Then it is easy to see that the curve $\beta_{n,m} : [0, s_{n,m}] \to \mathbf{R} \times N$ given by $\beta_{n,m}(s) = (-\frac{B_n - B_{n+m}}{s_{n,m}} s + B_n, \alpha_{n,m}(s))$ is a past-directed timelike curve with $\beta_{n,m}(0) = (B_n, y_n)$ and $\beta_{n,m}(s_{n,m}) = (B_{n+m}, y_{n+m})$. Also by Lemma 7.3.1, since the integral curve of ∇f passing through x_n realizes the distance between each level set

$\{B_{n+m}\} \times N$ and x_n for $m \geq 1$, we have

$$d(x_{n+m}, x_n) \leq d(\{B_{n+m}\} \times N, x_n) \leq d(\{0\} \times N, x_n) = B_n$$

for all $m \geq 1$. Thus $x_{n+m} \ll x_n$ for n, $m = 1, 2, \ldots$ and $d(x_{n+m}, x_n) \leq B_n$ for all $m \geq 1$. Then the timelike Cauchy completeness of $(M_{-1}, g_{-1}) = (\mathbf{R} \times N, -dt \otimes dt \oplus g_N)$ implies that $\{x_n\}$ has an accumulation point in $\mathbf{R} \times N$. Hence $\{y_n\}$ has an accumulation point in N.

Finally suppose that (M, g) satisfies Condition A. To show that (N, g_N) is complete, it suffices to show that every unit speed geodesic $\alpha : [0, b) \to N$ of (N, g_N) with $\alpha(0) = y \in N$ can be extended to $b = \infty$. Then it follows from Hopf-Rinow theorem that (N, g_N) is complete. Now suppose that α is inextendable beyond some $b < \infty$ and set $p = (0, y)$, $q = (\epsilon, y)$ and $\gamma(t) = (ct + \epsilon, \alpha(t))$ on $[0, b)$ in $\mathbf{R} \times N$, where $\epsilon > 0$ and $c > 1$. Then $p \ll q$ and γ is a future inextendable future-directed timelike geodesic starting at q. Now since $\gamma(t) \in \{ct + \epsilon\} \times N$ at each $t \in [0, b)$, $d(p, \gamma(t)) \leq ct + \epsilon < cb + \epsilon$ by Lemma 7.3.1 because the integral curve of ∇f passing through p realizes the distance between p and the level set $\{ct + \epsilon\} \times N$ for each $t \in [0, b)$. But since $(M_{-1}, g_{-1}) = (\mathbf{R} \times N,\ -dt \otimes dt \oplus g_N)$ satisfies Condition A, this gives a contradiction as $t \to b$. Thus $b = \infty$. □

Remark 7.3.3 Note that the main difference between the conclusions of Theorem 7.3.2 and Theorem 7.3.3 is that the Riemannian factor (N, g_N) is complete in the conclusion of Theorem 7.3.3. Also compare this with the conclusion of Theorem 7.3.1. This is an important difference since if (N, g_N) is a complete Riemannian manifold then $(M_{-1}, g_{-1}) = (\mathbf{R} \times N,\ -dt \otimes dt \oplus g_N)$ is a geodesically complete, globally hyperbolic spacetime. (See [BEE, p. 103]). Otherwise, if (N, g_N) is only a Riemannian manifold then $(M_{-1}, g_{-1}) = (\mathbf{R} \times N,\ -dt \otimes dt \oplus g_N)$ is a stably causal spacetime which is not necessarily geodesically complete.

The existence of solutions to timelike eikonal inequality and timelike eikonal equation has physical meanings in General Relativity. Recall that a future-directed unit timelike vector field Z on a spacetime (M, g) is called a *reference frame*. A reference frame Z on a spacetime (M, g) is called *synchronizable* (resp., *proper time synchronizable*) if there is a function $f : (M, g) \to \mathbf{R}$ such that $Z = -h\nabla f$ (resp., $Z = -\nabla f$), where $h = (-g(\nabla f, \nabla f))^{-1/2}$. Also, if $Z = -h\nabla f$ (resp., $Z = -\nabla f$) is a synchronizable (resp., proper time synchronizable) reference frame on a spacetime (M, g) then f is called a *time function* (resp., *proper time function*) for Z. Hence a spacetime (M, g) admits a synchronizable reference frame if and only if timelike eikonal inequality on (M, g) has a solution if and only if (M, g) is stably causal. Also a spacetime (M, g) admits a proper time synchronizable reference frame if and only if the timelike eikonal equation on (M, g) has a solution.

Now from the viewpoint of synchronizable and proper time synchronizable reference frames, we state the generalized semi-Riemannian regular interval theorem as a rigid singularity theorem for stably causal spacetimes. Recall that a spacetime (M, g) is called timelike geodesically complete if every timelike geodesic of (M, g) is complete.

Theorem 7.3.4 *Let (M, g) be a stably causal spacetime with a time function $f : (M, g) \to \mathbf{R}$ for a synchronizable reference frame Z. Also let $g_c = -g(\nabla f, \nabla f)g$. Then, either (M, g_c) is timelike geodesically incomplete or else, (M, g) is conformally diffeomorphic to a parametrized Lorentzian product manifold $(M_{-1}, g_{-1}) = (\mathbf{R} \times N, -dt \otimes dt \oplus g_{ct})$. In particular, if f is a proper time function for a proper time synchronizable reference frame Z on (M, g) then, either (M, g) is timelike geodesically incomplete or else, (M, g) is isometric to a parametrized Lorentzian product manifold $(M_{-1}, g_{-1}) = (\mathbf{R} \times N, -dt \otimes dt \oplus g_t)$.*

Proof. If (M, g_c) is timelike geodesically complete then since $\overset{c}{\nabla} f$ is a unit timelike geodesic vector field on (M, g_c) by Proposition 6.1.2, $\overset{c}{\nabla} f$ is complete. Hence by Theorem 6.3.2, (M, g) is conformally diffeomorphic to $(M_{-1}, g_{-1}) = (\mathbf{R} \times N, -dt \otimes dt \oplus g_{ct})$. The second claim follows similarly from Theorem 6.3.1. □

Further splittings of stably causal spacetimes into special cases of parametrized Lorentzian product manifolds also follow by relating the synchronizable reference frames to metric tensor.

Definition 7.3.2 A reference frame Z on a spacetime (M, g) is called *spatially conformally stationary* if there is a map $\mu : (M, g) \to (0, \infty)$ such that

$$(\mathcal{L}_{\mu Z} g)(X, Y) = 2\phi g(X, Y)$$

for every $X, Y \in \Gamma TM$ orthogonal to Z, where $\mathcal{L}$ is the Lie derivative and $\phi : (M, g) \to \mathbf{R}$ is a map. In particular, a spatially conformally stationary reference frame Z is called *spatially stationary* if $\phi = 0$.

Note that, if Z is spatially conformally stationary reference frame then also $(\mathcal{L}_Z g)(X, Y) = 2\phi' g(X, Y)$ for every $X, Y \in \Gamma TM$ orthogonal to Z, where $\phi' : (M, g) \to \mathbf{R}$ is a map. Indeed, if $(\mathcal{L}_{\mu Z} g)(X, Y) = 2\phi g(X, Y)$ for every X, $Y \in \Gamma TM$ orthogonal to Z then, since $(\mathcal{L}_{\mu Z} g)(X, Y) = \mu(\mathcal{L}_Z g)(X, Y)$ for every $X, Y \in \Gamma TM$ orthogonal to Z, it follows that $(\mathcal{L}_Z g)(X, Y) = 2\phi' g(X, Y)$ for every $X, Y \in \Gamma TM$ orthogonal to Z, where $\phi' = \frac{\phi}{\mu}$.

Remark 7.3.4 Let Z be a reference frame on an n-dimensional spacetime (M, g) and let $Z^\perp$ be the orthogonal bundle to Z. Then since $\nabla_X Z \in \Gamma Z^\perp$ for every $X \in \Gamma Z^\perp$, define a linear map

$$A_Z : \Gamma Z^\perp \to \Gamma Z^\perp$$

by

$$A_Z X = -\nabla_X Z.$$

Let $(A_Z)_{sym}$ and $(A_Z)_{sk}$ be the symmetric and the skew parts of A_Z, respectively. Note that $(A_Z)_{sym}$ further decomposes as

$$(A_Z)_{sym} = \frac{\operatorname{trace} A_Z}{n-1} id + ((A_Z)_{sym} - \frac{\operatorname{trace} A_Z}{n-1} id).$$

Hence by setting

$$\begin{aligned} \theta_Z &= \operatorname{trace} A_Z \ (\textit{expansion of } Z) \\ \sigma_Z &= (A_Z)_{sym} - \frac{\operatorname{trace} A_Z}{n-1} id \ (\textit{shear of } Z) \\ \xi_Z &= (A_Z)_{sk} \ (\textit{vorticity of } Z), \end{aligned}$$

A_Z has the irreducible decomposition

$$A_Z = \frac{\theta_Z}{n-1} id + \sigma_Z + \xi_Z.$$

Now let ω be a 1-form on M defined by

$$\omega = g^\flat(Z) = g(Z, \cdot),$$

where $g^\flat : \Gamma TM \to \Gamma TM^*$ is the musical isomorphism. Then it is easy to show that

$$g((A_Z)_{sym} X, Y) = -\frac{1}{2}(\mathcal{L}_Z g)(X, Y)$$

and

$$g((A_Z)_{sk} X, Y) = -d\omega(X, Y)$$

for every X, $Y \in \Gamma Z^\perp$. Hence, Z is spatially conformally stationary if and only if $\sigma_Z = 0$. In particular, Z is spatially stationary if and only if $(A_Z)_{sym} = 0$. Also it is easy to show that if Z is a synchronizable reference frame, that is, $Z = -h\nabla f$, then $d\omega(X, Y) = 0$ for every X, $Y \in \Gamma Z^\perp$ and hence $(A_Z)_{sk} = 0$. Thus, if Z is synchronizable then, Z is spatially conformally stationary if and only if $A_Z = \frac{\theta_Z}{n-1} id$, and Z is spatially stationary if and only if $A_Z = 0$.

Theorem 7.3.5 *Let (M, g) be a stably causal spacetime with a time function $f : (M, g) \to \mathbf{R}$ for a synchronizable reference frame $Z = -h\nabla f$, where $h = (-g(\nabla f, \nabla f))^{1/2}$. Also let $g_c = -g(\nabla f, \nabla f)g$. If Z is spatially conformally stationary then, either (M, g_c) is timelike geodesically incomplete, or else (M, g) is conformally diffeomorphic to a Lorentzian twisted product manifold $(M_{-1}, g_{-1}) = (\mathbf{R} \times N, -dt \otimes dt \oplus \lambda g_N)$.*

Proof. If (M, g_c) is timelike geodesically complete then, since $\overset{c}{\nabla} f$ is a unit timelike geodesic vector field on (M, g_c) by Proposition 6.1.2, $\overset{c}{\nabla} f$ is complete. Hence by Theorem 6.3.1, (M, g_c) is isometric to a parametrized Lorentzian product manifold $(\mathbf{R} \times f^{-1}(0), -dt \otimes dt \oplus g_{ct})$, where $g_{ct} = \Phi_t^*(g_{cf^{-1}(-t)})$. Hence to show that $g_{ct} = \lambda g_{c0}$, for all t, where $\lambda : \mathbf{R} \times f^{-1}(0) \to (0, \infty)$ is a function, it suffices to show that

$$\Phi_t\,|_{f^{-1}(0)}: (f^{-1}(0), g_{c0}) \to (f^{-1}(-t), g_{cf^{-1}(-t)})$$

is a conformal diffeomorphism for each t. Then it follows that $g_{ct} = \Phi_t^*(g_{cf^{-1}(-t)}) = \lambda_t g_{c0}$ for each t, where $\lambda_t : f^{-1}(0) \to (0, \infty)$ is a function, and hence λ_t defines a function $\lambda : \mathbf{R} \times f^{-1}(0) \to (0, \infty)$ by $\lambda(p, t) = \lambda_t(p)$. To show this, first note that, since $\nabla(-g(\nabla f, \nabla f)) = -2\nabla_{\nabla f}\nabla f$ by Corollary 3.1.1, it follows from Lemma 6.1.1-(3) that

$$H_f^c(X, Y) = H_f(X, Y) + \frac{H_f(\nabla f, \nabla f)}{g(\nabla f, \nabla f)} g(X, Y)$$

for every $X, Y \in \Gamma TM$ orthogonal to ∇f. Hence for every $X, Y \in \Gamma TM$ orthogonal to ∇f,

$$\begin{aligned}
(\mathcal{L}_{\overset{c}{\nabla} f} g_c)(X, Y) &= 2H_f^c(X, Y) \\
&= 2H_f(X, Y) + 2\tfrac{H_f(\nabla f, \nabla f)}{g(\nabla f, \nabla f)} g(X, Y) \\
&= (\mathcal{L}_{\nabla f} g)(X, Y) + 2\tfrac{H_f(\nabla f, \nabla f)}{g(\nabla f, \nabla f)} g(X, Y) \\
&= (\mathcal{L}_{(-\frac{1}{\mu h})\mu Z} g)(X, Y) - 2h^2 H_f(\nabla f, \nabla f) g(X, Y) \\
&= -\tfrac{1}{\mu h}(\mathcal{L}_{\mu Z} g)(X, Y) - 2h^2 H_f(\nabla f, \nabla f) g(X, Y) \\
&= -\tfrac{2\phi}{\mu h} g(X, Y) - 2h^2 H_f(\nabla f, \nabla f) g(X, Y) \\
&= -\tfrac{2}{h}(\tfrac{\phi}{\mu} + h^3 H_f(\nabla f, \nabla f)) g(X, Y).
\end{aligned}$$

Thus

$$\begin{aligned}(\mathcal{L}_{\overset{c}{\nabla} f} g_c)(X,Y) &= -2h(\tfrac{\phi}{\mu} + h^3 H_f(\nabla f, \nabla f)) g_c(X,Y) \\ &= 2k_c g_c(X,Y),\end{aligned}$$

where

$$k_c = -h(\frac{\phi}{\mu} + h^3 H_f(\nabla f, \nabla f)).$$

That is, $\overset{c}{\nabla} f$ is also a spatially conformally stationary reference frame on (M, g_c) up to minus or plus sign. Indeed this suffices to show that

$$\Phi_t \mid_{f^{-1}(0)}: (f^{-1}(0), g_{c0}) \to (f^{-1}(-t), g_{cf^{-1}(-t)})$$

is a conformal diffeomorphism for each t. To see this, let X_0, $Y_0 \in \Gamma T(f^{-1}(0))$ be nowhere vanishing local sections and let X, $Y \in \Gamma TM$ such that $X_{\Phi_t(p)} = (\Phi_t)_* X_0$, $Y_{\Phi_t(p)} = (\Phi_t)_* Y_0$. Then since $[\overset{c}{\nabla} f, X] = 0 = [\overset{c}{\nabla} f, Y]$,

$$\overset{c}{\nabla}_{\overset{c}{\nabla} f} X = \overset{c}{\nabla}_X \overset{c}{\nabla} f \quad \text{and} \quad \overset{c}{\nabla}_{\overset{c}{\nabla} f} Y = \overset{c}{\nabla}_Y \overset{c}{\nabla} f.$$

Now let

$$\cos_c(X,Y) = \frac{g_c(X,Y)}{g_c(X,X)^{1/2} g_c(Y,Y)^{1/2}}$$

be the cosine of the angle between the vector fields X and Y in (M, g_c). Then since

$$\begin{aligned}\tfrac{\partial}{\partial t} g_c((\Phi_t)_* X_0, (\Phi_t)_* Y_0) &= \tfrac{\partial}{\partial t} g_c(X \circ \Phi_t, Y \circ \Phi_t) \\ &= ((\overset{c}{\nabla} f) g_c(X,Y)) \circ \Phi_t \\ &= (g_c(\overset{c}{\nabla}_{\overset{c}{\nabla} f} X, Y) + g_c(X, \overset{c}{\nabla}_{\overset{c}{\nabla} f} Y)) \circ \Phi_t \\ &= (g_c(\overset{c}{\nabla}_X \overset{c}{\nabla} f, Y) + g_c(X, \overset{c}{\nabla}_Y \overset{c}{\nabla} f)) \circ \Phi_t \\ &= 2H_f^c(X,Y) \circ \Phi_t \\ &= ((\mathcal{L}_{\overset{c}{\nabla} f} g_c)(X,Y)) \circ \Phi_t \\ &= 2(k_c \circ \Phi_t) g_c(X,Y) \circ \Phi_t \\ &= 2(k_c \circ \Phi_t) g_c((\Phi_t)_* X_0, (\Phi_t)_* Y_0),\end{aligned}$$

it follows that

$$\frac{\partial}{\partial t}\cos_c((\Phi_t)_*X_0, (\Phi_t)_*Y_0)$$

$$= \frac{\partial}{\partial t}\left(\frac{g_c((\Phi_t)_*X_0, (\Phi_t)_*Y_0)}{g_c((\Phi_t)_*X_0, (\Phi_t)_*X_0)^{1/2} g_c((\Phi_t)_*Y_0, (\Phi_t)_*Y_0)^{1/2}}\right)$$

$$= \frac{2(k_c \circ \Phi_t) g_c((\Phi_t)_*X_0, (\Phi_t)_*Y_0)}{g_c((\Phi_t)_*X_0, (\Phi_t)_*X_0)^{1/2} g_c((\Phi_t)_*Y_0, (\Phi_t)_*Y_0)^{1/2}}$$

$$- \frac{2(k_c \circ \Phi_t) g_c((\Phi_t)_*X_0, (\Phi_t)_*Y_0)}{g_c((\Phi_t)_*X_0, (\Phi_t)_*X_0)^{1/2} g_c((\Phi_t)_*Y_0, (\Phi_t)_*Y_0)^{1/2}}$$

$$= 0.$$

Thus $\cos_c((\Phi_t)_*X_0, (\Phi_t)_*Y_0)$ is constant for all t and it follows that $\Phi_t \mid_{f^{-1}(0)}$ preserves the angles for each t. That is,

$$\Phi_t \mid_{f^{-1}(0)}: (f^{-1}(0), g_{c0}) \to (f^{-1}(-t), g_{cf^{-1}(-t)})$$

is a conformal diffeomorphism for each t. Hence (M, g_c) is isometric to the twisted Lorentzian product manifold $(M_{-1}, g_{-1}) = (\mathbf{R} \times N, -dt \otimes dt \oplus \lambda g_N)$, where $N = f^{-1}(0)$ and $g_N = g_{c0}$. Thus it follows that (M, g) is conformally diffeomorphic to a twisted Lorentzian product manifold $(M_{-1}, g_{-1}) = (\mathbf{R} \times N, -dt \otimes dt \oplus \lambda g_N)$. □

Remark 7.3.5 In Theorem 7.3.5, the twisting function $\lambda : \mathbf{R} \times N \to (0, \infty)$ can be determined in terms of k_c defined in the proof of Theorem 7.3.5. Note that, since (M, g_c) is isometric to twisted Lorentzian product manifold $(\mathbf{R} \times N, -dt \otimes dt \oplus \lambda g_N)$, where $N = f^{-1}(0)$ and $g_N = g_{c0}$, for $X_0, Y_0 \in \Gamma TN$,

$$g_c((\Phi_t)_*X_0, (\Phi_t)_*Y_0) = \lambda_t g_{c0}(X_0, Y_0),$$

where $\lambda_t : N \to (0, \infty)$ defined by $\lambda_t(p) = \lambda(t, p)$. Then since

$$\frac{\partial}{\partial t} g_c((\Phi_t)_*X_0, (\Phi_t)_*Y_0) = 2(k_c \circ \Phi_t) g_c((\Phi_t)_*X_0, (\Phi_t)_*Y_0),$$

we obtain,

$$\frac{\partial \lambda_t}{\partial t} g_{c0}(X_0, Y_0) = 2(k_c \circ \Phi_t)\lambda_t g_{c0}(X_0, Y_0).$$

Thus, by the initial condition

$$g_{c0} = \Phi_0^*(g_{cf^{-1}(0)}) = \lambda_0 g_{c0},$$

that is, $\lambda_0 = 1$,

$$\lambda_t(p) = e^{2\int_0^t (k_c \circ \Phi_s)(p) ds} = e^{2\int_0^t k_c(c_p(s)) ds},$$

where $c_p(s)$ is the integral curve of $\overset{c}{\nabla} f$ with $c_p(0) = p$. One can also determine k_c in terms of the function f. For, let $\{\overset{c}{\nabla} f, X_1^c, \ldots, X_{n-1}^c\}$ be an orthonormal local frame with respect to g_c, where $n = \dim M$. Then since

$$\begin{aligned}
\overset{c}{\Delta} f &= H_f^c(\overset{c}{\nabla} f, \overset{c}{\nabla} f) - \sum_{i=1}^{n-1} H_f^c(X_i^c, X_i^c) \\
&= -\sum_{i=1}^{n-1} H_f^c(X_i^c, X_i^c) \\
&= -\frac{1}{2}\sum_{i=1}^{n-1} (\mathcal{L}_{\overset{c}{\nabla} f} g_c)(X_i^c, X_i^c) \\
&= -\frac{1}{2}\sum_{i=1}^{n-1} 2k_c g_c(X_i^c, X_i^c) \\
&= -k_c \sum_{i=1}^{n-1} g_c(X_i^c, X_i^c) \\
&= -(n-1)k_c,
\end{aligned}$$

it follows that

$$k_c = -\frac{1}{n-1} \overset{c}{\Delta} f.$$

Also by Lemma 6.1.1–(4), since $\overset{c}{\Delta} f$ and Δf are related by

$$\overset{c}{\Delta} f = h^2 \Delta f + (n-2) h^4 H_f(\nabla f, \nabla f),$$

we have

$$k_c = -\frac{1}{n-1}\left(h^2 \Delta f + (n-2) h^4 H_f(\nabla f, \nabla f)\right).$$

Corollary 7.3.2 *Let (M, g) be a stably causal spacetime with a time function $f : (M, g) \to \mathbf{R}$ for a synchronizable reference frame $Z = -h\nabla f$, where $h = (-g(\nabla f, \nabla f))^{1/2}$. Also let $g_c = -g(\nabla f, \nabla f) g$ and $k_c = -\frac{1}{n-1}(h^2 \Delta f + (n-2) h^4 H_f(\nabla f, \nabla f))$ be a function on M, where $n = \dim M$. If Z is spatially conformally stationary then, either (M, g_c) is timelike geodesically incomplete, or else (M, g) is conformally diffeomorphic to*

(1) *a Lorentzian warped product manifold* $(M_{-1}, g_{-1}) = (\mathbf{R} \times N, -dt \otimes dt \oplus \lambda g_N)$, *provided that* k_c *is constant on each level surface (fiber) of* f,

(2) *a Lorentzian product manifold* $(M_{-1}, g_{-1}) = (\mathbf{R} \times N, -dt \otimes dt \oplus g_N)$, *provided that* $k_c = 0$ *on* M, *where* (N, g_N) *is a complete Riemannian manifold.*

Proof. First note that, by Theorem 7.3.5, if (M, g_c) is timelike geodesically complete then (M, g) is conformally diffeomorphic to a Lorentzian twisted product manifold $(M_{-1}, g_{-1}) = (\mathbf{R} \times N, -dt \otimes dt \oplus \lambda g_N)$. Also by Remark 7.3.5, the twisting function $\lambda_t(p) = \lambda(t, p)$ is given by

$$\lambda_t(p) = e^{2 \int_0^t (k_c \circ \Phi_s)(p) ds} = e^{2 \int_0^t k_c(c_p(s)) ds},$$

where $c_p(s)$ is the integral curve of $\overset{c}{\nabla} f$ with $c_p(0) = p$.

(1) Note that, if k_c is constant on each level surface (fiber) of f then λ_t is constant at each $t \in \mathbf{R}$. Thus λ is only a function of t and hence (M_{-1}, g_{-1}) is a Lorentzian warped product manifold.

(2) Note that if $k_c = 0$ on M then λ_t is equal to 1 for every $t \in \mathbf{R}$. Thus λ is the constant function 1 and hence (M_{-1}, g_{-1}) is a Lorentzian product manifold. Then also note that, since (M_{-1}, g_{-1}) is timelike geodesically complete Lorentzian product manifold, (N, g_N) is a complete Riemannian manifold. □

Corollary 7.3.3 *Let* (M, g) *be a stably causal spacetime with a time function* $f : (M, g) \to \mathbf{R}$ *for a synchronizable reference frame* $Z = -h\nabla f$, *where* $h = (-g(\nabla f, \nabla f))^{1/2}$. *Also let* $g_c = -g(\nabla f, \nabla f) g$ *and* $k_c = -h^4 H_f(\nabla f, \nabla f)$ *be a function on* M. *If* Z *is spatially stationary then, either* (M, g_c) *is timelike geodesically incomplete, or else* (M, g) *is conformally diffeomorphic to*

(1) *a Lorentzian warped product manifold* $(M_{-1}, g_{-1}) = (\mathbf{R} \times N, -dt \otimes dt \oplus \lambda g_N)$, *provided that* k_c *is constant on each level surface (fiber) of* f,

(2) *a Lorentzian product manifold* $(M_{-1}, g_{-1}) = (\mathbf{R} \times N, -dt \otimes dt \oplus g_N)$, *provided that* $k_c = 0$ *on* M, *where* (N, g_N) *is a complete Riemannian manifold.*

Proof. Since Z is spatially stationary, the expression of k_c in the proof of Theorem 7.3.5 becomes

$$k_c = -h^4 H_f(\nabla f, \nabla f).$$

Hence the claim follows from Corollary 7.3.2. □

Corollary 7.3.4 *Let (M, g) be a spacetime with a proper time function $f : (M, g) \to \mathbf{R}$ for a proper time synchronizable reference frame $Z = -\nabla f$. If Z is spatially stationary then, either (M, g) is timelike geodesically incomplete, or else (M, g) is isometric to a Lorentzian product manifold $(M_{-1}, g_{-1}) = (\mathbf{R} \times N, \ -dt \otimes dt \oplus g_N)$, where (N, g_N) is a complete Riemannian manifold.*

Proof. Since Z is a geodesic vector field by Proposition 6.1.2, $H_f(\nabla f, \nabla f) = 0$. Hence $k_c = -h^4 H_f(\nabla f, \nabla f) = 0$ in the statement of Corollary 7.3.3. Also since $-g(\nabla f, \nabla f) = 1$, $g_c = g$ in the statement of Corollary 7.3.3. Thus by Corollary 7.3.3–(2), either (M, g) is timelike geodesically incomplete or else (M, g) is conformally diffeomorphic to a Lorentzian product manifold $(M_{-1}, g_{-1}) = (\mathbf{R} \times N, \ -dt \otimes dt \oplus g_N)$, where (N, g_N) is a complete Riemannian manifold. But since $g_c = g$, it follows that (M, g) is isometric to (M_{-1}, g_{-1}). (See the proof of Theorem 6.3.2). □

Appendix A

Submanifolds of Semi-Riemannian Manifolds

Let (M, g) be a semi-Riemannian manifold. Let H be a submanifold of (M, g), $i : H \to M$ the inclusion map, and g_H $(= i^*g)$ be the restriction of g to H. If g_H is a metric tensor on H (of index ν') then (H, g_H) is called a *semi-Riemannian submanifold* of (M, g) (of index ν'). If g_H is of constant type (μ', ν', η') at every $p \in M$ with $\mu' \geq 1$, then (H, g_H) is called a *degenerate submanifold* of (M, g) (of type (μ', ν', η')).

In this appendix, we briefly review some properties of semi-Riemannian and degenerate submanifolds of semi-Riemannian manifolds encountered in this book. We devote the first section to a review of certain properties of semi-Riemannian submanifolds of semi-Riemannian manifolds. The second section of this appendix reviews some properties of degenerate submanifolds of semi-Riemannian manifolds. The interested reader is referred to [K2], [DB] for detailed study of degenerate submanifolds.

Throughout this appendix, let (M, g) be a semi-Riemannian manifold of index ν and dimension n.

A.1 Semi-Riemannian Submanifolds

Let (M, g) be a semi-Riemannian manifold with the Levi-Civita connection ∇. Let H be a submanifold of M and $i : H \to M$ be the inclusion map. We also denote the pullback of ∇ along i by ∇. We call a vector field along i a *vector field along* H. Also since $i : H \to M$ is an embedding which is the identity onto its image, $Y \in \Gamma TH$ can be identified with $i_*Y \in \Gamma_i Ti(H)$. Thus, by abusing notation, for X, $Y \in \Gamma TH$, we also write $\nabla_X Y$ for $\nabla_X i_* Y$.

Let (H, g_H) be a semi-Riemannian submanifold of (M, g). Then since (T_pM, g_p)

is a nondegenerate subspace of (T_pM, g_p) at each $p \in M$, we have a canonical orthogonal direct sum decomposition of T_pM as

$$T_pM = T_pH \oplus T_pH^{\perp}$$

at each $p \in H$, where $T_pH^{\perp}$ is the orthogonal space to T_pH in (T_pM, g_p) which is also nondegenerate. (See Proposition 1.1.1). Thus every vector $x \in T_pM$, $p \in H$, can be uniquely written as

$$x = x^T + x^{\perp},$$

where x^T and $x^{\perp}$ are the components of x in T_pH and $T_pH^{\perp}$, respectively. Consequently, $TM\mid_H$ has a canonical orthogonal direct sum decomposition

$$TM\mid_H = TH \oplus TH^{\perp},$$

where $TH^{\perp}$ is the orthogonal bundle to TH given by $TH^{\perp} = \bigcup_{p\in H} T_pH^{\perp}$. Hence if X is a vector field along H then X can be uniquely written as

$$X = X^T + X^{\perp},$$

where X^T and $X^{\perp}$ are the components of X in ΓTH and $\Gamma TH^{\perp}$, respectively. In fact, this property of semi-Riemannian submanifolds makes the study of semi-Riemannian submanifolds of semi-Riemannian manifolds quite similar to the study of submanifolds of Riemannian manifolds.

Note that, if (H, g_H) is a semi-Riemannian submanifold of (M, g) then, since (H, g_H) is intrinsically a semi-Riemannian manifold, (H, g_H) intrinsically has the Levi-Civita connection D by Theorem 2.2.2. In fact, it can easily be observed from Theorem 2.2.3 and the uniqueness of the Levi-Civita connection of (H, g_H) that we have the following relation between D and ∇. (Also see [O, p. 99]).

Proposition A.1.1 *Let (H, g_H) be a semi-Riemannian submanifold of (M, g). Then for every X, $Y \in \Gamma TH$,*

$$D_XY = (\nabla_XY)^T.$$

Thus, if (H, g_H) is a semi-Riemannian submanifold of (M, g), then for every X, $Y \in \Gamma TH$,

$$\nabla_XY = D_XY + (\nabla_XY)^{\perp}.$$

Here $(\nabla_XY)^{\perp}$ is something new.

Definition A.1.1 Let (H, g_H) be a semi-Riemannian submanifold of (M, g). Then the map

$$\mathrm{I} : \Gamma TH \times \Gamma TH \to \Gamma TH^{\perp}$$

defined by

$$\mathrm{I}(X, Y) = (\nabla_X Y)^{\perp}$$

is called the *second fundamental form tensor* of (H, g_H).

Remark A.1.1 It is easy to show that I is linear in its arguments and symmetric. Thus, it can be shown as in Proposition 2.1.1 that the value of $\mathrm{I}(X, Y)$ at $p \in H$ only depends on the values of X and Y at p. Thus one can define $\mathrm{I}(x, y)$ by

$$\mathrm{I}(x, y) = \mathrm{I}(X, Y)(p),$$

where X, $Y \in \Gamma TH$ with $X(p) = x$ and $Y(p) = y$.

Hence, if (H, g_H) is a semi-Riemannian submanifold of (M, g) then, for every X, $Y \in \Gamma TH$,

$$\nabla_X Y = D_X Y + \mathrm{I}(X, Y).$$

This decomposition leads to the famous Gauss equation for (H, g_H).

Theorem A.1.1 (The Gauss Equation). *Let (H, g_H) be a semi-Riemannian submanifold of (M, g). Also let R_H and R be the curvature tensors of (H, g_H) and (M, g), respectively. Then for every X, Y, Z, $V \in \Gamma TH$,*

$$\begin{aligned} g(R(X,Y)Z,V) &= g(R_H(X,Y)Z,V) \\ &\quad + g(\mathrm{I}(X,Z),\mathrm{I}(Y,V)) - g(\mathrm{I}(Z,Y),\mathrm{I}(X,V)). \end{aligned}$$

Proof. See, for example, [O, p. 100]. □

Now we define an important class of submanifolds of semi-Riemannian manifolds.

Definition A.1.2 Let H be a submanifold of (M, g). Then H is called *totally geodesic* if $\nabla_X Y \in \Gamma TH$ for every X, $Y \in \Gamma TH$.

Remark A.1.2 Note that, if (H, g_H) is a semi-Riemannian submanifold of (M, g) then, H is totally geodesic if and only if second fundamental form tensor I of (H, g_H) is vanishing.

First we make the following observation on totally geodesic submanifolds of a semi-Riemannian manifold.

Proposition A.1.2 *Let H be a connected, totally geodesic submanifold of (M, g). Then (H, g_H) is either a semi-Riemannian or a degenerate submanifold of (M, g).*

Proof. First we show that the parallel translation of a vector tangent to H along a curve in H remains tangent to H. Note that, it suffices to show this property locally, that is, for every injective portion of this curve. For let $\gamma : I \to H$ be an injective curve with $\gamma(t_0) = p \in H$, where $t_0 \in I \subseteq \mathbf{R}$ is an open interval. Let X_1, ..., $X_m \in \Gamma_\gamma TH$ be such that $\{X_1(t), \ldots, X_m(t)\}$ forms a basis for $T_{\gamma(t)}H$ at each $t \in I$, where $m = \dim H$. Also let $X_{m+1}, \ldots, X_n \in \Gamma_\gamma TM$ such that $\{X_1(t), \ldots, X_m(t), X_{m+1}(t), \ldots, X_n(t)\}$ forms a basis for $T_{\gamma(t)}M$ at each $t \in I$. Let $\{\tilde{X}_1, \ldots, \tilde{X}_m\}$ be an extension of $\{X_1, \ldots, X_m\}$ to a neighborhood U of $\gamma(t_0) = p$ in M such that $\tilde{X}_i(q) \in T_qH$, $i = 1, \ldots, m$, for each $q \in H \cap U$. (Note that, without loss of generality we may assume that U contains the image of γ). Now let $x \in T_pH$ and X be a parallel vector field along γ with $X(t_0) = x$. Then

$$X = \sum_{i=1}^{n} h^i X_i,$$

where $h^i : I \to \mathbf{R}$ with $h^i(t_0) = 0$ for $i = m+1, \ldots, n$. Thus,

$$\begin{aligned}
0 = \nabla_{\frac{d}{dt}} X &= \sum_{i=1}^{n} \left(\frac{dh^i}{dt} X_i + h^i \nabla_{\frac{d}{dt}} X_i \right) \\
&= \sum_{i=1}^{m} \left(\frac{dh^i}{dt} X_i + h^i \nabla_{\frac{d}{dt}} (\tilde{X}_i \circ \gamma) \right) \\
&\quad + \sum_{i=m+1}^{n} \left(\frac{dh^i}{dt} X_i + h^i \nabla_{\frac{d}{dt}} X_i \right) \\
&= \sum_{i=1}^{m} \left(\frac{dh^i}{dt} X_i + h^i \nabla_{\dot{\gamma}} \tilde{X}_i \right) \\
&\quad + \sum_{i=m+1}^{n} \left(\frac{dh^i}{dt} X_i + h^i \nabla_{\frac{d}{dt}} X_i \right).
\end{aligned}$$

Now, since H is totally geodesic, $\nabla_{\dot{\gamma}} \tilde{X}_i$ is tangent to H at each $\gamma(t)$ and it follows that

$$\sum_{i=m+1}^{n} \left(\frac{dh^i}{dt} X_i + h^i \nabla_{\frac{d}{dt}} X_i \right)$$

is tangent to H for each $t \in I$. Thus if we write

$$\nabla_{\frac{d}{dt}} X_i = \sum_{j=1}^{n} k_i^j X_j, \quad i = m+1, \ldots, n,$$

where $k_i^j : I \to \mathbf{R}$, then

$$\begin{aligned}
&\sum_{i=m+1}^{n} \left(\frac{dh^i}{dt} X_i + h^i \nabla_{\frac{d}{dt}} X_i \right) \\
&= \sum_{i=m+1}^{n} \left(\frac{dh^i}{dt} X_i + h^i \sum_{j=1}^{n} k_i^j X_j \right) \\
&= \sum_{i=m+1}^{n} \frac{dh^i}{dt} X_i + \sum_{i=1}^{m} \left(\sum_{j=m+1}^{n} h^j k_j^i \right) X_i \\
&\quad + \sum_{i=m+1}^{n} \left(\sum_{j=m+1}^{n} h^j k_j^i \right) X_i \\
&= \sum_{i=1}^{m} \left(\sum_{j=m+1}^{n} h^j k_j^i \right) X_i + \sum_{i=m+1}^{n} \left(\frac{dh^i}{dt} + \sum_{j=m+1}^{n} h^j k_j^i \right) X_i
\end{aligned}$$

is tangent to H for each $t \in I$. Hence

$$\sum_{i=m+1}^{n} \left(\frac{dh^i}{dt} + \sum_{j=m+1}^{n} h^j k_j^i \right) X_i = 0$$

and, since $\{X_{m+1}, \ldots, X_n\}$ is a linearly independent set,

$$\frac{dh^i}{dt} + \sum_{j=m+1}^{n} h^j k_j^i = 0, \quad i = m+1, \ldots, n.$$

But this is a homogeneous first order linear differential system satisfying the initial conditions

$$h^i(t_0) = 0 \quad \text{for} \quad i = m+1, \ldots, n.$$

Thus by the uniqueness of its solution,

$$h^i(t) = 0 \quad \text{for all} \quad t \in I,$$

where $i = m+1, \ldots, n$. Hence $X = \sum_{i=1}^{m} h^i X_i$, that is, X is tangent to H at each $t \in I$.

Now let $p, q \in H$ and $\gamma : I \to H$ be a curve with $\gamma(t_1) = p$ and $\gamma(t_2) = q$, where $t_1, t_2 \in I$. Let $\{x_1, \ldots, x_m\}$ be an orthonormal basis for T_pH with respect to g_{H_p}. Then since the parallel translation $P_{t_1}^{t_2}(\gamma) : (T_pM, g_p) \to (T_qM, g_q)$ is an isometry and parallel translation of a vector tangent to H along a curve in H remains tangent to H, $\{P_{t_1}^{t_2}(\gamma)(x_1), \ldots, P_{t_1}^{t_2}(\gamma)(x_m)\}$ is also an orthonormal basis for T_qH with respect to g_{H_q}. Thus g_{H_p} and g_{H_q} is of the same type and it follows that (H, g_H) is either a semi-Riemannian or a degenerate submanifold of (M, g). □

In fact, if (H, g_H) is a semi-Riemannian submanifold of (M, g) then we have the following equivalent definitions for (H, g_H) to be totally geodesic.

Proposition A.1.3 *Let (H, g_H) be a semi-Riemannian submanifold of (M, g). Then the following are equivalent:*

(1) *H is totally geodesic.*

(2) $\mathrm{I} = 0$.

(3) *Every geodesic of (H, g_H) is also a geodesic of (M, g).*

(4) *If $v \in T_pM$ is tangent to H then the geodesic γ_v of (M, g) lies initially in H.*

(5) *If α is a curve in H and $v \in T_{\alpha(t_0)}H$ then parallel translation of v along α is the same for (H, g_H) and for (M, g).*

Proof. See, for example, [O, p. 104]. □

Definition A.1.3 Let (H, g_H) be a semi-Riemannian submanifold of (M, g). Then the *mean curvature vector field* $\mathcal{N}$ of (H, g_H) is defined to be the trace of I with respect to g_H, that is,

$$\begin{aligned} \mathcal{N} &= \sum_{i=1}^{m} g_H(X_i, X_i)\mathrm{I}(X_i, X_i) \\ &= \sum_{i=1}^{m} g(X_i, X_i)\mathrm{I}(X_i, X_i), \end{aligned}$$

where $m = \dim H$ and $\{X_1, \ldots, X_m\}$ is an orthonormal local frame.

Another important class of semi-Riemannian submanifolds is the critical semi-Riemannian submanifolds.

Definition A.1.4 Let (H, g_H) be a semi-Riemannian submanifold of (M, g). Then (H, g_H) is called *critical* if $\mathcal{N} = 0$.

Observe from Propositions 6.2.3 and 6.2.4 that a semi-Riemannian submersion $f : (M, g) \to (\mathbf{R}, kdt \otimes dt)$, where k is either of -1 or 1, is affine if and only if fibers of f are totally geodesic. Next we will show such a relation between harmonic maps and critical semi-Riemannian submanifolds.

Proposition A.1.4 *Let $f : (M, g) \to (\mathbf{R}, kdt \otimes dt)$ be a semi-Riemannian submersion with fibers as semi-Riemannian submanifolds of (M, g), where k is either of -1 or 1. Then, f is harmonic on (M, g) if and only if the fibers of f are critical submanifolds of (M, g).*

Proof. Let $\{X_1, \ldots, X_{n-1}\}$ be an orthonormal local frame for ker f_*, where $n = \dim M$. Then note that $\{\nabla f, X_1, \ldots, X_{n-1}\}$ is an orthonormal local frame for TM by Proposition 6.1.1–(1). Hence since ∇f is a geodesic vector field by Proposition 6.1.2,

$$\begin{aligned}\Delta f &= -\operatorname{div} \nabla f \\ &= -\sum_{i=1}^{n-1} g(X_i, X_i) g(\nabla_{X_i} \nabla f, X_i) \\ &= \sum_{i=1}^{n-1} g(X_i, X_i) g(\nabla f, \nabla_{X_i} X_i) \\ &= \sum_{i=1}^{n-1} g(X_i, X_i) g(\nabla f, \mathrm{I}(X_i, X_i)) \\ &= g(\nabla f, \mathcal{N}).\end{aligned}$$

Thus $\Delta f = 0$ if and only if $\mathcal{N} = 0$, that is, f is harmonic on (M, g) if and only if fibers of f are critical. □

For a detailed study of critical semi-Riemannian submanifolds of a semi-Riemannian manifold, see [K5].

A.2 Degenerate Submanifolds

Let (H, g_H) be a degenerate submanifold of (M, g) of type (μ', ν', η'). Then (T_pH, g_{H_p}) is a degenerate subspace of (T_pM, g_p) of type (μ', ν', η') at every $p \in H$. Let

L_{gH_p} be the degenerate space of g_{H_p} in T_pH. Hence $L_{gH_p} = T_pH \cap T_pH^\perp$ and $L^\perp_{gH_p} = T_pH + T_pH^\perp$ at each $p \in H$, where $T_pH^\perp$ is the orthogonal space to T_pH. (See Remark 1.1.2 and Proposition 1.1.3). But since (T_pH, g_{H_p}) is a degenerate subspace of (T_pM, g_p),

$$T_pM \neq T_pH + T_pH^\perp = L^\perp_{gH_p}$$

at each $p \in H$. (See Proposition 1.1.1). In fact, this is the major problem in the study of the geometry of degenerate submanifolds. Indeed, if we set

$$L_{gH} = \bigcup_{p \in H} L_{gH_p},$$

since $L_{gH} = TH \cap TH^\perp$, where $TH^\perp$ is the orthogonal bundle to TH defined by $TH^\perp = \bigcup_{p\in H} T_pH^\perp$, and dim $L_{gH_p} = \mu'$ (constant) at every $p \in H$, L_{gH} is a subbundle of TH and $TH^\perp$, which we call the *degenerate subbundle* of (H, g_H). Thus $L^\perp_{gH} = TH + TH^\perp$ is a vector bundle over H, where $L^\perp_{gH}$ is the orthogonal bundle to L_{gH} defined by $L^\perp_{gH} = \bigcup_{p\in H} L^\perp_{gH_p}$. But since (H, g_H) is a degenerate submanifold of (M, g),

$$TM\,|_H \neq TH + TH^\perp = L^\perp_{gH}.$$

Thus in general, if $X, Y \in \Gamma TH$ then $\nabla_X Y \notin \Gamma L^\perp_{gH}$, and hence $\nabla_X Y$ cannot be written as a sum of a vector field in ΓTH and a vector field in $\Gamma TH^\perp$. Also note that, even if $\nabla_X Y \in \Gamma L^\perp_{gH}$, there is no canonical way of choosing components of $\nabla_X Y$ in ΓTH and $\Gamma TH^\perp$. But this problem can be taken care of by several means as we will see. Thus the study of semi-Riemannian submanifolds cannot be carried to degenerate submanifolds in a straightforward way. Therefore, first we will investigate the conditions on degenerate submanifolds that insure $\nabla_X Y \in \Gamma L^\perp_{gH}$ for every $X, Y \in \Gamma TH$, which is the most essential requirement in the study of degenerate submanifolds. As one expects, these conditions on (H, g_H) are related to L_{gH}.

Definition A.2.1 Let (H, g_H) be a degenerate submanifold of (M, g). Then (H, g_H) is called *integrable* if L_{gH} is an integrable subbundle of TH.

Proposition A.2.1 *Let (H, g_H) be a degenerate submanifold of (M, g). Then the following are equivalent:*

(1) *(H, g_H) is integrable.*

(2) *$\nabla_U U' \in \Gamma TH^\perp$ for every $U, U' \in \Gamma L_{gH}$.*

(3) *$\nabla_U X \in \Gamma L^\perp_{gH}$ for every $U \in \Gamma L_{gH}$, $X \in \Gamma TH$.*

(4) $\nabla_X U \in \Gamma L_{g_H}^{\perp}$ *for every* $X \in \Gamma TH$, $U \in \Gamma L_{g_H}$.

Proof. (1) $\Longleftrightarrow$ (2): Let $X \in \Gamma TH$ and U, $U' \in \Gamma L_{g_H}$. Then since $[U, U']$, $[X, U]$, $[X, U'] \in \Gamma TH$,

$$\begin{aligned} g([U,U'],X) &= g(\nabla_U U', X) - g(\nabla_{U'} U, X) \\ &= g(\nabla_U U', X) + g(U, \nabla_{U'} X) \\ &= g(\nabla_U U', X) + g(U, \nabla_X U') \\ &= g(\nabla_U U', X) - g(\nabla_X U, U') \\ &= g(\nabla_U U', X) - g(\nabla_U X, U') \\ &= g(\nabla_U U', X) + g(X, \nabla_U U') \\ &= 2g(\nabla_U U', X). \end{aligned}$$

Therefore $[U, U'] \in \Gamma L_{g_H}$ if and only if $\nabla_U U' \in \Gamma TH^{\perp}$.

(2) $\Longleftrightarrow$ (3) $\Longleftrightarrow$ (4): Let $X \in \Gamma TH$ and U, $U' \in \Gamma L_{g_H}$. Since $[U, X] \in \Gamma TH$,

$$\begin{aligned} g(\nabla_U U', X) &= -g(U', \nabla_U X) \\ &= -g(U', \nabla_X U). \end{aligned}$$

Hence $\nabla_U U' \in \Gamma TH^{\perp}$ if and only if $\nabla_U X \in \Gamma L_{g_H}^{\perp}$ if and only if $\nabla_X U \in \Gamma L_{g_H}^{\perp}$. □

Definition A.2.2 Let (H, g_H) be a degenerate submanifold of (M, g). Then (H, g_H) is called *irrotational* if $\nabla_X U \in \Gamma TH$ for every $X \in \Gamma TH$, $U \in \Gamma L_{g_H}$.

Note that, if (H, g_H) is an irrotational degenerate submanifold of (M, g) then since $TH \subseteq L_{g_H}^{\perp}$, (H, g_H) is integrable by Proposition A.2.1-(4).

Proposition A.2.2 *Let* (H, g_H) *be a degenerate submanifold of* (M, g). *Then the following are equivalent:*

(1) (H, g_H) *is irrotational.*

(2) $\nabla_X Z \in \Gamma L_{g_H}^{\perp}$, *for every* $X \in \Gamma TH$, $Z \in \Gamma TH^{\perp}$.

(3) $\nabla_U Z \in \Gamma TH^{\perp}$, *for every* $U \in \Gamma L_{g_H}$, $Z \in \Gamma TH^{\perp}$.

(4) $\nabla_U X \in \Gamma TH$, *for every* $U \in \Gamma L_{g_H}$, $X \in \Gamma TH$.

Proof. (1) $\Longleftrightarrow$ (2): Let $U \in \Gamma L_{g_H}$, $X \in \Gamma TH$ and $Z \in \Gamma TH^{\perp}$. Then since $g(U, Z) = 0$,

$$g(\nabla_X U, Z) = -g(U, \nabla_X Z).$$

Hence $\nabla_X U \in \Gamma TH$ if and only if $\nabla_X Z \in \Gamma L_{g_H}^{\perp}$.

(2) $\Longleftrightarrow$ (3): Let $U \in \Gamma L_{g_H}$, $X \in \Gamma TH$ and $Z \in \Gamma TH^{\perp}$. Then since

$$g(X, Z) = g([U, X], Z) = g(U, Z) = 0,$$

we obtain

$$\begin{aligned} g(X, \nabla_U Z) &= -g(\nabla_U X, Z) \\ &= -g(\nabla_X U, Z) \\ &= g(U, \nabla_X Z). \end{aligned}$$

Hence $\nabla_U Z \in \Gamma TH^{\perp}$ if and only if $\nabla_X Z \in \Gamma L_{g_H}^{\perp}$.

(4) $\Longleftrightarrow$ (1): Let $U \in \Gamma L_{g_H}$ and $X \in \Gamma TH$. Then since $[U, X] \in \Gamma TH$ and $\nabla_U X = \nabla_X U + [U, X]$, $\nabla_U X \in \Gamma TH$ if and only if $\nabla_X U \in \Gamma TH$. □

Note that if (H, g_H) is a degenerate hypersurface in (M, g) then $L_{g_H} = TH^{\perp}$ and (H, g_H) is necessarily of type $(1, \nu-1, n-\nu-1)$ by Proposition 1.1.4. Hence a degenerate hypersurface (H, g_H) is integrable and it can be easily seen from Proposition A.2.1–(4) that (H, g_H) is irrotational.

Definition A.2.3 Let (H, g_H) be a degenerate submanifold of (M, g). Then (H, g_H) is called *stationary* if $\mathcal{L}_U g_H = 0$ on H for every $U \in \Gamma L_{g_H}$, where $\mathcal{L}$ is the Lie derivative on H.

Proposition A.2.3 *Let (H, g_H) be a degenerate submanifold of (M, g). Then the following are equivalent:*

(1) *(H, g_H) is stationary.*

(2) *$\nabla_X U \in \Gamma TH^{\perp}$ for every $X \in \Gamma TH$, $U \in \Gamma L_{g_H}$.*

(3) *$\nabla_X Y \in \Gamma L_{g_H}^{\perp}$ for every X, $Y \in \Gamma TH$.*

Proof. Let $X, Y \in \Gamma TH$ and $U \in \Gamma L_{g_H}$.

(1) $\Longleftrightarrow$ (2): First note that

$$
\begin{aligned}
g(\nabla_X U, Y) &= Xg(U,Y) - g(U, \nabla_X Y) \\
&= -g(U, \nabla_Y X + [X,Y]) \\
&= -g(U, \nabla_Y X) \\
&= -Yg(U,X) + g(\nabla_Y U, X) \\
&= g(\nabla_Y U, X) = g(X, \nabla_Y U).
\end{aligned}
$$

Hence,

$$
\begin{aligned}
(\mathcal{L}_U g_H)(X,Y) &= Ug_H(X,Y) - g_H(\mathcal{L}_U X, Y) - g_H(X, \mathcal{L}_U Y) \\
&= Ug(X,Y) - g(\mathcal{L}_U X, Y) - g(X, \mathcal{L}_U Y) \\
&= g(\nabla_U X, Y) + g(X, \nabla_U Y) - g(\nabla_U X, Y) \\
&\quad + g(\nabla_X U, Y) - g(X, \nabla_U Y) + g(X, \nabla_Y U) \\
&= g(\nabla_X U, Y) + g(X, \nabla_Y U) \\
&= 2g(\nabla_X U, Y).
\end{aligned}
$$

Hence $\mathcal{L}_U g_H = 0$ if and only if $\nabla_X U \in \Gamma TH^\perp$.

(2) $\Longleftrightarrow$ (3): Since $g(\nabla_X Y, U) = -g(Y, \nabla_X U)$, $\nabla_X Y \in \Gamma L_{g_H}^\perp$ if and only if $\nabla_X U \in \Gamma TH^\perp$. □

Notice that, it follows by Proposition A.2.3–(2) and Proposition A.2.1 that every stationary degenerate submanifold (H, g_H) of (M,g) is integrable. Also, it can easily be seen from Proposition A.2.3–(3) that every stationary degenerate hypersurface (H, g_H) in (M,g) is totally geodesic.

Proposition A.2.4 *Let (H, g_H) be a degenerate submanifold of (M,g). Then the following are equivalent:*

(1) *(H, g_H) is irrotational and stationary.*

(2) *$\nabla_X U \in \Gamma L_{g_H}$ for every $X \in \Gamma TH$, $U \in \Gamma L_{g_H}$.*

(3) *$\nabla_X Z \in \Gamma L_{g_H}^\perp$ for every $X \in \Gamma TH$, $Z \in \Gamma L_{g_H}^\perp$.*

Proof. (1) $\Longleftrightarrow$ (2): Follows immediately from Definition A.2.2 and Proposition A.2.3–(2) since $L_{g_H} = TH \cap TH^\perp$.

(2) $\iff$ (3): Let $U \in \Gamma L_{g_H}$. Then since $g(\nabla_X Z, U) = -g(Z, \nabla_X U)$, it follows that $\nabla_X Z \in \Gamma L_{g_H}^{\perp}$ if and only if $\nabla_X U \in \Gamma L_{g_H}$. □

Now returning to Chapter 5, recall that a semi-Riemannian transversal map $f : (M_1, g_1) \to (M_2, g_2)$ to a semi-Riemannian foliation $(\mathcal{F}, g_{2\mathcal{F}})$ on (M_2, g_2) is called *regular* if $A_1 = (\ker f_*^{\perp}) + (\ker f_*^{\perp})^{\perp}$ is a totally geodesic distribution. Now we show that if $f : (M_1, g_1) \to (M_2, g_2)$ is a regular semi-Riemannian transversal map to a semi-Riemannian foliation $(\mathcal{F}, g_{2\mathcal{F}})$ on (M_2, g_2) then the leaves of $f^*(\mathcal{F})$ are irrotational, stationary submanifolds of (M_1, g_1), provided that they are degenerate. (See Remark 5.2.4).

Proposition A.2.5 *Let $f : (M_1, g_1) \to (M_2, g_2)$ be a regular semi-Riemannian transversal map to a semi-Riemannian foliation $(\mathcal{F}, g_{2\mathcal{F}})$ on (M_2, g_2), where $\mathcal{F} = \{S_\alpha\}_{\alpha \in A}$. If $\{(f^{-1}(S_\alpha), g_{1f^{-1}(S_\alpha)})\}_{\alpha \in A}$ are degenerate submanifolds of (M_1, g_1) then each $(f^{-1}(S_\alpha), g_{1f^{-1}(S_\alpha)})$, $\alpha \in A$, is an irrotational, stationary degenerate submanifold of (M_1, g_1).*

Proof. First note that, each $f^{-1}(S_\alpha)$ is an integral manifold of $\ker f_*^{\perp}$. Hence if we set $(H_\alpha, g_{1H_\alpha}) = (f^{-1}(S_\alpha), g_{1f^{-1}(S_\alpha)})$ for each $\alpha \in A$ then $TH_\alpha = (\ker f_*^{\perp}) |_{H_\alpha}$, $TH_\alpha^{\perp} = (\ker f_*^{\perp})^{\perp} |_{H_\alpha}$, $L_{g_1 H_\alpha} = L_1 |_{H_\alpha}$ and $L_{g_1 H_\alpha}^{\perp} = A_1 |_{H_\alpha}$. Now since A_1 is totally geodesic, $\nabla_X Z \in \Gamma A_1$ for every $X \in \Gamma \ker f_*^{\perp}$ and $Z \in \Gamma A_1$. Thus $\nabla_X Z \in \Gamma L_{g_1 H_\alpha}^{\perp}$ for every $X \in \Gamma TH_\alpha$, $Z \in \Gamma L_{g_1 H_\alpha}^{\perp}$ and it follows from Proposition A.2.4–(3) that each $(H_\alpha, g_{1H_\alpha})$ is an irrotational, stationary degenerate submanifold of (M_1, g_1). □

Notice that Proposition A.2.4–(3) gives us a hope to define the second fundamental form tensor of a stationary degenerate submanifold.

Definition A.2.4 Let (H, g_H) be a degenerate submanifold of (M, g). Then

(1) $(\overline{TH}, \overline{g}_H)$ is called the *nondegenerate quotient tangent bundle* of (H, g), where $\overline{TH} = TH/L_{g_H} = \bigcup_{p \in H} T_pH/L_{g_{H_p}}$ and $\overline{g}_H(\overline{X}, \overline{Y}) = g(X, Y)$, where $X, Y \in \Gamma TH$ with $\pi_{TH}(X) = \overline{X}$, $\pi_{TH}(Y) = \overline{Y}$ and $\pi_{TH} : TH \to \overline{TH}$ is the natural projection.

(2) $(\overline{TH}^{\perp}, \overline{g}_{\overline{TH}^{\perp}})$ is called the *nondegenerate quotient orthogonal bundle* of (H, g), where $\overline{TH}^{\perp} = TH^{\perp}/L_{g_H} = \bigcup_{p \in H} T_pH^{\perp}/L_{g_{H_p}}$ and $\overline{g}_{\overline{TH}^{\perp}}(\overline{X}, \overline{Y}) = g(X, Y)$, where $X, Y \in \Gamma TH^{\perp}$ with $\pi_{TH^{\perp}}(X) = \overline{X}$, $\pi_{TH^{\perp}}(Y) = \overline{Y}$ and $\pi_{TH^{\perp}} : TH^{\perp} \to \overline{TH}^{\perp}$ is the natural projection.

(3) $(\overline{L^{\perp}_{g_H}}, \overline{g}_{\overline{L^{\perp}_{g_H}}})$ is called the *nondegenerate quotient total bundle* of (H,g), where $\overline{L^{\perp}_{g_H}} = L^{\perp}_{g_H}/L_{g_H} = \bigcup_{p\in H} L^{\perp}_{g_{H_p}}/L_{g_{H_p}}$ and $\overline{g}_{\overline{L^{\perp}_{g_H}}}(\overline{X},\overline{Y}) = g(X,Y)$, where X, $Y \in \Gamma L^{\perp}_{g_H}$ with $\pi_{L^{\perp}_{g_H}}(X) = \overline{X}$, $\pi_{L^{\perp}_{g_H}}(Y) = \overline{Y}$ and $\pi_{L^{\perp}_{g_H}} : L^{\perp}_{g_H} \to \overline{L^{\perp}_{g_H}}$ is the natural projection.

Note that $(\overline{TH}, \overline{g}_H)$, $(\overline{TH}^{\perp}, \overline{g}_{\overline{TH}^{\perp}})$ and $(\overline{L^{\perp}_{g_H}}, \overline{g}_{\overline{L^{\perp}_{g_H}}})$ are semi-Riemannian vector bundles over H and $\overline{L^{\perp}_{g_H}}$ has the canonical orthogonal direct sum decomposition

$$\overline{L^{\perp}_{g_H}} = \overline{TH} \oplus \overline{TH}^{\perp}.$$

(See Remark 1.1.3).

Now Proposition A.2.4–(3) enables us to define the second fundamental form tensor of a stationary degenerate submanifold.

Definition A.2.5 Let (H, g_H) be a stationary degenerate submanifold of (M, g). Then the map

$$\mathbf{I} : \Gamma TH \times \Gamma TH \to \Gamma\overline{TH}^{\perp}$$

defined by

$$\mathbf{I}(X,Y) = (\pi_{L^{\perp}_{g_H}}(\nabla_X Y))^{\perp}$$

is called the *second fundamental form tensor* of (H, g_H), where $(\pi_{L^{\perp}_{g_H}}(\nabla_X Y))^{\perp}$ is the component of $\pi_{L^{\perp}_{g_H}}(\nabla_X Y)$ in $\Gamma\overline{TH}^{\perp}$.

Note that by Proposition A.2.4–(3), $\mathbf{I}$ is well-defined. Also it is easy to show that $\mathbf{I}$ is linear in its arguments and symmetric. Thus it can be shown as in Proposition 2.1.1 that the value of $\mathbf{I}(X,Y)$ at $p \in H$ only depends on the values of X and Y at p. Hence one can define $\mathbf{I}(x,y)$ by

$$\mathbf{I}(x,y) = \mathbf{I}(X,Y)(p),$$

where X, $Y \in \Gamma TH$ with $X(p) = x$ and $Y(p) = y$.

By the definition of $\mathbf{I}$, the following proposition is obvious.

Proposition A.2.6 *Let (H, g_H) be a stationary degenerate submanifold of (M, g). Then, (H, g_H) is totally geodesic if and only if $\mathbf{I} = 0$.*

The next thing is to obtain the Gauss equation for a degenerate submanifold. But for this, first we have to introduce an intrinsic geometric structure to a degenerate submanifold. Observe that a degenerate submanifold (H, g_H) is intrinsically a manifold H with a symmetric $(0,2)$-tensor field g_H of type (μ', ν', η') at every $p \in M$. This observation motivates us to define singular semi-Riemannian manifolds.

Definition A.2.6 Let H be a manifold and h be a symmetric $(0,2)$-tensor field on H of constant type (μ, ν, η) at every $p \in M$, where $\mu \geq 1$. Then (H, h) is called a *singular semi-Riemannian manifold.*

Let (H, h) be a singular semi-Riemannian manifold. Define the *degenerate subbundle* of (H, h) by

$$L_h = \bigcup_{p \in H} L_{h_p},$$

where L_{h_p} is the degenerate space of h_p. As in the case of degenerate submanifolds, we will call (H, h) *stationary* if $\mathcal{L}_U h = 0$ for every $U \in \Gamma L_h$, where $\mathcal{L}$ is the Lie derivative. Also define the *nondegenerate quotient tangent bundle* $(\overline{TH}, \overline{h})$ of (H, h) by $\overline{TH} = TH/L_h = \bigcup_{p \in H} T_pH/L_{h_p}$ and $\overline{h}(\overline{X}, \overline{Y}) = h(X, Y)$, where X, $Y \in \Gamma TH$ with $\pi(X) = \overline{X}$, $\pi(Y) = \overline{Y}$ and $\pi : TH \to \overline{TH}$ is the natural projection. Note that $(\overline{TH}, \overline{h})$ is a semi-Riemannian vector bundle over M.

From viewpoint of vector bundles, a connection on a manifold is in fact, a connection in its tangent bundle. The concept of a connection on a manifold can be generalized to a concept of a connection in a vector bundle as follows:

Definition A.2.7 Let E be a real vector bundle over a manifold H. A map $\nabla : \Gamma TH \times \Gamma E \to \Gamma E$ is called a *connection* in E, if given X, $Y \in \Gamma TH$ and U, $V \in \Gamma E$,

(1) $\nabla_{X+Y} U = \nabla_X U + \nabla_Y U$,

(2) $\nabla_{kX} U = k \nabla_X U$, $k \in C^\infty(H)$,

(3) $\nabla_X (U + V) = \nabla_X U + \nabla_X V$,

(4) $\nabla_X (kU) = X(k)U + k\nabla_X U$, $k \in C^\infty(H)$.

Also if ∇ is a connection in a vector bundle E over a manifold H, the curvature tensor R of ∇ is defined by

$$R(X, Y)U = \nabla_X \nabla_Y U - \nabla_Y \nabla_X U - \nabla_{[X,Y]} U,$$

where X, $Y \in \Gamma TH$, $U \in \Gamma E$ and $[\,,\,]$ is the Lie bracket. It is easy to check that the map

$$R : \Gamma TH \times \Gamma TH \times \Gamma E \to \Gamma E$$

defined by

$$R(X, Y, U) = R(X, Y)U$$

is linear in its arguments. Thus it can be shown as in Proposition 2.1.1 that the value of $R(X, Y)U$ at $p \in H$ only depends on the values of X, Y and U at p. Hence one can define $R(x, y)u$ by

$$R(x, y)u = (R(X, Y)U)(p),$$

where X, $Y \in \Gamma TH$, $U \in \Gamma E$ with $X(p) = x$, $Y(p) = y$, $U(p) = u$.

The following theorem shows that, if $(H,\, h)$ is a stationary singular semi-Riemannian manifold then the induced singular semi-Riemannian structure on H by h defines a canonical connection in $\overline{TH}$.

Theorem A.2.1 *Let (H, h) be a stationary singular semi-Riemannian manifold. Then there exists a unique connection $\overline{\nabla}$ in $\overline{TH}$ such that*

(1) $Z\overline{h}(\overline{X}, \overline{Y}) = \overline{h}(\overline{\nabla}_Z \overline{X}, \overline{Y}) + \overline{h}(\overline{X}, \overline{\nabla}_Z \overline{Y})$

(2) $\overline{\nabla}_X \overline{Y} - \overline{\nabla}_Y \overline{X} = \pi([X, Y])$,

where X, Y, $Z \in \Gamma TH$ with $\pi(X) = \overline{X}$, $\pi(Y) = \overline{Y}$.

Proof. See [K2, p. 47]. □

Definition A.2.8 Let (H, h) be a stationary singular semi-Riemannian manifold. Then the unique connection $\overline{\nabla}$ in $\overline{TH}$ which is given by Theorem A.2.1 is called the *Koszul connection* of (H, h).

Note here that the curvature tensor $\overline{R}$ of the Koszul connection of a stationary singular semi-Riemannian manifold satisfies identities similar to the curvature tensor of the Levi-Civita connection of a semi-Riemannian manifold.

Proposition A.2.7 *Let $\overline{\nabla}$ be the Koszul connection of a stationary singular semi-Riemannian manifold (H, h) and let $\overline{R}$ be the curvature tensor of $\overline{\nabla}$. Then for all X, Y, Z, $V \in \Gamma TH$ with $\pi(X) = \overline{X}$, $\pi(Y) = \overline{Y}$, $\pi(Z) = \overline{Z}$, $\pi(V) = \overline{V}$,*

(1) $\overline{R}(X,Y)\overline{Z} = -\overline{R}(Y,X)\overline{Z}$,

(2) $\overline{h}(\overline{R}(X,Y)\overline{Z},\overline{V}) = -\overline{h}(\overline{R}(X,Y)\overline{V},\overline{Z})$,

(3) $\overline{R}(X,Y)\overline{Z} + \overline{R}(Y,Z)\overline{X} + \overline{R}(Z,X)\overline{Y} = 0$,

(4) $\overline{h}(\overline{R}(X,Y)\overline{Z},\overline{V}) = \overline{h}(\overline{R}(Z,V)\overline{X},\overline{Y})$.

Proof. See [K2, p. 49]. □

Now let (H, g_H) be a stationary degenerate submanifold of (M, g). Then since (H, g_H) is intrinsically a stationary singular semi-Riemannian manifold, by Theorem A.2.1, (H, g_H) intrinsically has the Koszul connection $\overline{\nabla}$. In fact, the Koszul connection $\overline{\nabla}$ of (H, g_H) is related to the Levi-Civita connection ∇ of (M, g).

Proposition A.2.8 *Let (H, g_H) be a stationary degenerate submanifold of (M, g). Then for every X, $Y \in \Gamma TH$ with $\pi_{TH}(Y) = \overline{Y}$,*

$$\overline{\nabla}_X \overline{Y} = (\pi_{L^{\perp}_{g_H}}(\nabla_X Y))^T,$$

where $(\pi_{L^{\perp}_{g_H}}(\nabla_X Y))^T$ is the component of $\pi_{L^{\perp}_{g_H}}(\nabla_X Y)$ in $\Gamma\overline{TH}$.

Proof. See [K2, p. 73]. □

Hence note that, if (H, g_H) is a stationary degenerate submanifold of (M, g) then for every X, $Y \in \Gamma TH$ with $\pi_{TH}(Y) = \overline{Y}$,

$$\pi_{L^{\perp}_{g_H}}(\nabla_X Y) = \overline{\nabla}_X \overline{Y} + \mathsf{I}(X,Y).$$

This decomposition leads to the Gauss equation for an irrotational, stationary degenerate submanifold (H, g_H).

Theorem A.2.2 (The Gauss Equation) *Let (H, g_H) be an irrotational, stationary degenerate submanifold of (M, g). Also let $\overline{R}_H$ be the curvature tensor of the Koszul connection $\overline{\nabla}$ of (H, g_H) and R be the curvature tensor of (M, g). Then for every X, Y, Z, $V \in \Gamma TH$ with $\pi_{TH}(Z) = \overline{Z}$, $\pi_{TH}(V) = \overline{V}$,*

$$\begin{aligned} g(R(X,Y)Z,V) &= \overline{g}_H(\overline{R}_H(X,Y)\overline{Z},\overline{V}) \\ &\quad + \overline{g}_{\overline{TH}^{\perp}}(\mathsf{I}(X,Z),\mathsf{I}(Y,V)) + \overline{g}_{\overline{TH}^{\perp}}(\mathsf{I}(Z,Y),\mathsf{I}(X,V)) \end{aligned}$$

Proof. See [K2, p. 75]. □

In Theorem A.2.2, (H, g_H) is assumed to be irrotational and stationary rather than (H, g_H) being only stationary. Note that, if (H, g_H) is only stationary then $\nabla_X Y \in L_{g_H}^{\perp}$ for every X, $Y \in \Gamma TH$. But in general, $\nabla_X \nabla_Y Z \notin \Gamma L_{g_H}^{\perp}$ for every X, Y, $Z \in \Gamma TH$. However note that $\nabla_X \nabla_Y Z \in \Gamma L_{g_H}^{\perp}$ for every X, Y, $Z \in \Gamma TH$ if (H, g_H) is irrotational and stationary by Proposition A.2.4–(3). In fact, this suffices for $R(X,Y)Z \in \Gamma L_{g_H}^{\perp}$ for every X, Y, $Z \in \Gamma TH$ which enables one to relate R_H and $\overline{R}$ via $\mathbf{I}$. For a detailed study of degenerate submanifolds, see [K2], [DB].

Appendix B

Riemannian and Lorentzian Geometry

In this appendix, we review some global concepts in Riemannian and Lorentzian geometry encountered in this book. We devote the first section of this appendix to a fundamental concept of Riemannian geometry, that is, the Riemannian distance function. We devote the second section of this appendix to some fundamental concepts in Lorentzian geometry, that is, the causal structure of Lorentzian manifolds and the Lorentzian distance function. The concepts we cover in this appendix are very fundamental and should be well-known to experienced Riemannian and Lorentzian geometers.

B.1 Riemannian Geometry

Throughout this section let (M, g) be a Riemannian manifold of dimension n.

The most important property of the Riemannian structure of (M, g) is the Riemannian distance function

$$d : M \times M \to [0, \infty)$$

defined on (M, g). Given $p, q \in M$, let Ω_{pq} be the set of all piecewise smooth curves in M from p to q. Given a curve $\gamma \in \Omega_{pq}$ with $\gamma : [a, b] \to M$, there is a finite partition $a = t_0 < t_1 < \cdots < t_k = b$ such that $\gamma \mid (t_i, t_{i+1})$ is smooth for each i. The *length* of γ in (M, g) is defined as

$$L(\gamma) = \sum_{i=0}^{k-1} \int_{t_i}^{t_{i+1}} g(\dot{\gamma}(t), \dot{\gamma}(t))^{1/2} dt.$$

Definition B.1.1 Let (M, g) be a connected Riemannian manifold and p, $q \in M$. Then the *Riemannian distance* $d(p, q)$ *from* p *to* q is defined by

$$d(p, q) = \inf \{L(\gamma)/\gamma \in \Omega_{pq}\}.$$

For any Riemannian manifold (M, g), the function $d : M \times M \to \mathbf{R}$ has the following properties:

Proposition B.1.1 *Let* (M, g) *be a connected Riemannian manifold. Then the function* $d : M \times M \to \mathbf{R}$ *is a metric on* M*, that is, for all* p, q, $r \in M$,

(1) $d(p, q) \geq 0$, *and* $d(p, q) = 0$ *if and only if* $p = q$,

(2) $d(p, q) = d(q, p)$,

(3) $d(p, q) \leq d(p, r) + d(r, q)$.

Furthermore,

(4) $d : M \times M \to [0, \infty)$ *is continuous and the family of metric balls*

$$B(p, \varepsilon) = \{q \in M/d(p, q) < \varepsilon\}$$

for all $\varepsilon > 0$ *forms a basis for the given manifold topology.*

Proof. See, for example [O, p. 136]. □

By the definition of Riemannian distance, a curve $\gamma \in \Omega_{pq}$ with $d(p, q) = L(\gamma)$ is the shortest curve from p to q. In this case, we call γ *minimal.*

Remark B.1.1 Note that, if $\gamma : [a, b] \to M$ in Ω_{pq} is minimal, then for all s, t with $a \leq s, t \leq b$, we have $d(\gamma(s), \gamma(t)) = L(\gamma \mid [s, t])$. Motivating from this observation, we will call a curve $\gamma : I \to M$, where I is an interval, *minimal* if $d(\gamma(s), \gamma(t)) = L(\gamma \mid [s, t])$ for all s, $t \in I$.

Corollary B.1.1 *Let* (M, g) *be a connected Riemannian manifold. If* $\gamma \in \Omega_{pq}$ *is minimal then* γ *can be reparametrized to be a (smooth) geodesic from* p *to* q.

Proof. See, for example, [O, p. 137]. □

Remark B.1.2 Note that if $\gamma : I \to M$ is a minimal curve, then by Corollary B.1.1, γ is a geodesic. Also an inextendable minimal geodesic of a Riemannian manifold is called a *line.*

Note that if (M, g) is a Riemannian manifold then (M, d) is a metric space by Proposition B.1.1. The following famous theorem of Riemannian geometry states a relation between completeness of (M, d) as a metric space and geodesic completeness of (M, g) as a Riemannian manifold. (See Definition 2.2.8).

Theorem B.1.1 (Hopf-Rinow) *Let (M, g) be a connected Riemannian manifold. Then the following are equivalent:*

(1) *Metric completeness: (M, d) is a complete metric space.*

(2) *Geodesic completeness: (M, g) is geodesically complete.*

(3) *For some $p \in M$, the exponential map $\exp_p$ is defined on T_pM.*

(4) *Finite compactness: Every subset K of M that is d-bounded has compact closure.*

Furthermore, if anyone of (1) *through* (4) *holds then,*

(5) *Minimal geodesic connectibility: Given $p, q \in M$, there exists a smooth geodesic γ from p to q with $L(\gamma) = d(p, q)$.*

Proof. See, for example, [O, p. 138]. □

A connected Riemannian manifold (M, g) is called *complete* if anyone (and hence all) of conditions (1) to (4) is satisfied by (M, g). Also note that, any compact connected Riemannian manifold is complete.

B.2 Lorentzian Geometry

Throughout this section, let (M, g) be a Lorentzian manifold of dimension n.

A Lorentzian manifold does not necessarily admit a timelike vector field. If (M, g) admits a timelike vector field Z then (M, g) is said to be *time oriented* by Z. The timelike vector field Z divides all causal tangent vectors into two separate classes, called future and past-directed. A causal tangent vector $v \in T_pM$ is called *future* (resp., *past*) *directed* if $g(Z(p), v) < 0$ (resp., $g(Z(p), v) > 0$). A Lorentzian manifold (M, g) is called *time orientable* if (M, g) admits a time orientation by some timelike vector field Z.

Definition B.2.1 A connected, time oriented Lorentzian manifold (M, g) is called a *spacetime.*

Let (M, g) be a spacetime. A curve $\gamma : I \to M$ is called *timelike* (resp., *null*) if $\dot{\gamma}(t)$ is timelike (resp., null) for all $t \in I$. Thus a timelike or null curve $\gamma : I \to M$ is either future or past-directed, that is, either $g((Z \circ \gamma)(t), \dot{\gamma}(t)) < 0$ or $g((Z \circ \gamma)(t), \dot{\gamma}(t)) > 0$ for all $t \in I$, respectively, where Z is a time orientation for (M, g). A curve $\gamma : I \to M$ is called *causal* if $\dot{\gamma}(t)$ is either timelike or null at each $t \in I$. Thus, a causal curve is also either future or past-directed. We use the notation $p \ll q$ for p, $q \in M$ if there is a future-directed timelike curve from p to q, and $p \leq q$ for p, $q \in M$ if either $p = q$ or there is a future-directed causal curve from p to q. Furthermore, $p < q$ for p, $q \in M$ means there is a future-directed causal curve from p to q.

Note that the relations $\ll$ and $\leq$ are clearly transitive. Moreover

$$p \ll q \quad \text{and} \quad q \leq r \quad \text{implies} \quad p \ll r$$

and

$$p \leq q \quad \text{and} \quad q \ll r \quad \text{implies} \quad p \ll r.$$

(Cf. [Pe, p. 13]).

Definition B.2.2 Let (M, g) be a spacetime. For a given $p \in M$, the *chronological future* $I^+(p)$, *chronological past* $I^-(p)$, *causal future* $J^+(p)$ and *causal past* $J^-(p)$ of p are defined as follows:

$$I^+(p) = \{q \in M / p \ll q\}$$

$$I^-(p) = \{q \in M / q \ll p\}$$

$$J^+(p) = \{q \in M / p \leq q\}$$

$$J^-(p) = \{q \in M / q \leq p\}.$$

Proposition B.2.1 *Let (M, g) be a spacetime. Then for any $p \in M$, $I^+(p)$ and $I^-(p)$ are open subsets of M.*

Proof. See [Pe, p. 13]. □

Note that, in general, $J^+(p)$ or $J^-(p)$ may not be closed subsets of M.

Now we review some causality conditions encountered in this book.

A spacetime (M, g) which does not contain any closed timelike curve, that is, $p \notin I^+(p)$ for all $p \in M$, is called *chronological*. A spacetime which does not contain any closed causal curve is called *causal*, that is, a causal spacetime contains no pair of distinct points p and q with $p \leq q \leq p$.

Proposition B.2.2 *Any compact spacetime (M, g) contains a closed timelike curve and thus fails to be chronological.*

Proof. See, for example, [Pe, p. 38]. □

An open subset U of a spacetime (M, g) is called *causally convex* if no timelike curve intersects U more than once. A spacetime (M, g) is called *strongly causal at* $p \in M$ if there is a neighborhood base at p consisting of causally convex sets. A spacetime (M, g) is called *strongly causal* if it is strongly causal at every $p \in M$.

A strongly causal spacetime (M, g) can be characterized in terms of the Alexandrov topology for M. The *Alexandrov topology* on a spacetime (M, g) is the topology given M by taking as a basis all sets of the form $I^+(p) \cap I^-(q)$ for all p, $q \in M$. The given manifold topology on M is always finer than the Alexandrov topology since $I^+(p) \cap I^-(q)$ is an open set in the given topology by Proposition B.2.1.

Proposition B.2.3 *Let (M, g) be a spacetime. The Alexandrov topology for (M, g) agrees with the given manifold topology if and only if (M, g) is strongly causal.*

Proof. See, for example, [Pe, p. 34]. □

Let (M, g) be a spacetime. A Lorentzian metric tensor g_0 on M is called *wider* than g if $g_0(v, v) < 0$ for every nonzero $v \in TM$ with $g(v, v) \leq 0$. (Note that if (M, g) is time oriented by Z then (M, g_0) is also time oriented by Z). For a given spacetime (M, g), there always exists a Lorentzian metric tensor g_0 on M wider than g. Indeed, let Z be a timelike vector field on (M, g) and let ω be a 1-form on (M, g) defined by $\omega = g(Z, \cdot)$. Then it is easy to show that $g_0 = g - \omega \otimes \omega$ is a Lorentzian metric tensor on M wider than g. A spacetime (M, g) is called *stably causal* if there exists a spacetime (M, g_0) which is chronological with g_0 wider than g.

A stably causal spacetime (M, g) can be characterized in terms of the existence of a function on M with timelike gradient everywhere.

Proposition B.2.4 *Let (M, g) be a spacetime. Then, (M, g) is stably causal if and only if there is a function $f : M \to \mathbf{R}$ such that ∇f is everywhere timelike.*

Proof. In one direction the proof is simple. Suppose that such an f exists and let $\omega = df = g(\nabla f, \cdot)$. Then

$$g_0 = g - \omega \otimes \omega$$

is a Lorentzian metric tensor on M wider than g. We show that (M, g_0) is chronological which suffices to show that (M, g) is stably causal. Let $\overset{0}{\nabla} f$ be the gradient of f on (M, g_0). First we show that $\overset{0}{\nabla} f$ is timelike with respect to g_0. Note that, since

$$g_0(\overset{0}{\nabla} f, \overset{0}{\nabla} f) = df(\overset{0}{\nabla} f) = g(\nabla f, \overset{0}{\nabla} f),$$

it follows that

$$\begin{aligned} g(\nabla f, \nabla f) &= df(\nabla f) \\ &= g_0(\overset{0}{\nabla} f, \nabla f) \\ &= g(\overset{0}{\nabla} f, \nabla f) - g(\nabla f, \overset{0}{\nabla} f) g(\nabla f, \nabla f) \\ &= g_0(\overset{0}{\nabla} f, \overset{0}{\nabla} f) - g_0(\overset{0}{\nabla} f, \overset{0}{\nabla} f) g(\nabla f, \nabla f) \\ &= g_0(\overset{0}{\nabla} f, \overset{0}{\nabla} f)(1 - g(\nabla f, \nabla f)). \end{aligned}$$

Hence since $g(\nabla f, \nabla f) < 0$ it follows that $g_0(\overset{0}{\nabla} f, \overset{0}{\nabla} f) < 0$, that is, $\overset{0}{\nabla} f$ is also timelike. Now let $\gamma : I \to M$ be a timelike curve in (M, g_0). Then since

$$\frac{d}{dt}(f \circ \gamma)(t) = g_0(((\overset{0}{\nabla} f) \circ \gamma)(t), \dot{\gamma}(t)) \neq 0$$

for every $t \in I$, it follows that f is either increasing or decreasing along every timelike curve in (M, g_0). Hence (M, g_0) does not contain any closed timelike curve.

For the idea of the (harder) proof of the converse, see [HE, p. 198]. □

Note that, in the terminology used in this book, a spacetime (M, g) is stably causal if and only if the timelike eikonal inequality has a solution.

Remark B.2.1 Let (M, g) be a spacetime. A continuous function $f : (M, g) \to \mathbf{R}$ is called a *global time function* on (M, g) if f is increasing along each future-directed causal curve. In fact the existence of a global time function on (M, g) is also sufficient for (M, g) to be stably causal. (Cf. [Ha]).

A strongly causal spacetime (M, g) is called *globally hyperbolic* if for each pair of points p, $q \in M$, $J^+(p) \cap J^-(q)$ is compact.

A globally hyperbolic spacetime can be characterized using Cauchy hypersurfaces. A *Cauchy hypersurface* S in a spacetime (M, g) is a subset of M such that every inextendable causal curve in (M, g) intersects S exactly once.

Proposition B.2.5 *Let (M, g) be a spacetime. Then (M, g) is globally hyperbolic if and only if (M, g) admits a Cauchy hypersurface.*

Proof. See [HE, p. 211]. □

In this section, we introduced the causality conditions in an order of increasing strength. That is

globally hyperbolic

$\Downarrow$

stably causal

$\Downarrow$

stongly causal

$\Downarrow$

causal

$\Downarrow$

chronological.

Since the proofs of these implications are not available in the standard literature in Lorentzian geometry, we provide their proofs here.

First note that causal $\Longrightarrow$ chronological is obvious. Now we show that strongly causal $\Longrightarrow$ causal.

Proposition B.2.6 *A strongly causal spacetime (M, g) is causal.*

Proof. Suppose there exists a closed causal curve γ in (M, g). Assume that γ is future-directed. Let $p \in$ image (γ) and U be a causally convex set containing p such that γ is not contained in U. Let $x, y \in U$ such that $x \ll p$ and $p \ll y$. Then image $(\gamma) \subseteq I^+(x)$ and image $(\gamma) \subseteq I^-(y)$. Now let $q \in$ image (γ) with $q \notin U$. Thus since $q \in I^+(x)$ and $q \in I^-(y)$, there are future-directed timelike curves α_1 and α_2 from x to q and q to y, respectively. Hence by combining α_1 and α_2, we obtain a future-directed timelike curve intersecting U more than once in contradiction to the strong causality. Thus (M, g) contains no closed causal curves. □

Next we show that stable causality $\Longrightarrow$ strong causality. For this, we need the following well-known fact about strong causality violation at a point.

Lemma B.2.1 *Let (M, g) be a spacetime. Then strong causality fails at $p \in M$ if and only if there exists $q \in M$ with $q < p$ such that $x \ll p$ and $q \ll y$ together imply $x \ll y$ for all x, y.*

Proof. See [Pe, p. 31]. □

Proposition B.2.7 *A stably causal spacetime (M, g) is strongly causal.*

Proof. Suppose (M, g) is not strongly causal at $p \in M$. Then by Lemma B.2.1, there exists $q \in M$ with $q < p$ such that $x \ll p$ and $q \ll y$ together imply $x \ll y$ for all x, y. Let $\{x_n\}$ and $\{y_n\}$ be a sequence of points with $x_n \ll p$ and $q \ll y_n$ such that $x_n \to p$ and $y_n \to q$ as $n \to \infty$. Then there exists a future-directed timelike curve γ_n from x_n to y_n for each n. Also let α be a future-directed causal curve from q to p and $r \in$ image (α) with $q < r < p$. Now let g_0 be a Lorentzian metric tensor on M wider than g and let $I_0^+(r)$ and $I_0^-(r)$ be the chronological future and past of r in (M, g_0). Since g_0 is wider than g, $I_0^+(r)$ and $I_0^-(r)$ are open neighborhoods of p and q, respectively. Consequently, $I_0^+(r)$ and $I_0^-(r)$ contain $\{x_n\}$ and $\{y_n\}$ for $n \geq N$ for some $N > 0$, respectively. Hence there is a future-directed timelike curve β_n from y_n to x_n in (M, g_0) for $n \geq N$. Thus by combining γ_n and β_n for each $n \geq N$, we obtain closed future-directed timelike curves in (M, g_0) in contradiction to stable causality. Thus (M, g) is strongly causal. □

Finally we show that globally hyperbolic $\Longrightarrow$ stable causality. For this, we need the following well-known fact about globally hyperbolic spacetimes.

Theorem B.2.1 *Let (M, g) be a globally hyperbolic spacetime. Then M is homeomorphic to $\mathbf{R} \times S$, where S is a topological hypersurface in M. Furthermore, if $\psi : \mathbf{R} \times S \to M$ is the homeomorphism, we can arrange it so that $\psi(t, S)$ is a Cauchy hypersurface for each t and $\psi(\mathbf{R}, s)$ is a timelike curve for each s.*

Proof. See [Pe, p. 48]. □

Proposition B.2.8 *A globally hyperbolic spacetime (M, g) is stably causal.*

Proof. By Theorem B.2.1, there is a homeomorphism $\psi : \mathbf{R} \times S \to M$ such that $\psi(t, S)$ is a Cauchy hypersurface for each t and $\psi(\mathbf{R}, s)$ is a timelike curve for each s. Let ψ^{-1} be the inverse of ψ and $\pi_1 : \mathbf{R} \times S \to \mathbf{R}$ be the projection. Then it is easy to see that the function $f : M \to \mathbf{R}$ defined by

$$f = \pi_1 \circ \psi^{-1}$$

is a global time function on (M, g) up to minus or plus sign. Hence by Remark B.2.1, (M, g) is stably causal. □

A subset H of a spacetime (M, g) is called *achronal* if the relation $p \ll q$ never holds for p, $q \in H$, that is, no timelike curve meets H more than once. Also a subset

H of a spacetime (M, g) is called *acausal* if the relation $p < q$ never holds for p, $q \in H$, that is, no causal curve meets H more than once. Note that acausality is stronger than achronality.

A hypersurface H in a spacetime (M, g) is called *spacelike* if (H, g_H) is a Riemannian manifold.

Proposition B.2.9 *Let (M, g) be a spacetime. Then an achronal spacelike hypersurface in (M, g) is acausal.*

Proof. See [O, p. 425]. □

An important property of the Lorentzian structure of a spacetime (M, g) is the Lorentzian distance function

$$d : M \times M \to [0, \infty) \cup \{\infty\}$$

defined on (M, g).

Let (M, g) be a spacetime. Given p, $q \in M$, let Ω_{pq} be the set of all piecewise smooth future-directed causal curves from p to q. Given a curve $\gamma \in \Omega_{pq}$ with $\gamma : [a, b] \to M$, there is a finite partition $a = t_0 < t_1 < \cdots < t_k = b$ such that $\gamma \mid (t_i, t_{i+1})$ is a smooth future-directed causal curve for each i. Then the length of γ in (M, g) is defined as

$$L(\gamma) = \sum_{i=0}^{k-1} \int_{t_i}^{t_{i+1}} (-g(\dot{\gamma}(t), \dot{\gamma}(t)))^{1/2} dt.$$

Definition B.2.3 Let (M, g) be a spacetime and p, $q \in M$. Then the *Lorentzian distance $d(p, q)$ from p to q* is defined by

$$d(p, q) = \begin{cases} \sup \{L(\gamma)/\gamma \in \Omega_{pq}\} & \text{if} \quad \Omega_{pq} \neq \emptyset \\ 0 & \text{if} \quad \Omega_{pq} = \emptyset. \end{cases}$$

Clearly the Lorentzian distance function d is not symmetric. Also $d(p, q)$ may not be finite. But for any spacetime, the Lorentzian distance function $d : M \times M \to [0, \infty) \cup \{\infty\}$ has the following properties:

Proposition B.2.10 *Let (M, g) be a spacetime. Then*

(1) $d(p, q) > 0$ if and only if $p \ll q$.

(2) If $p \ll q \ll r$ then $d(p,r) \geq d(p,q) + d(q,r)$.

(3) $d : M \times M \to [0,\infty) \cup \{\infty\}$ is lower semicontinuous.

Proof. See [O, p. 410]. □

For globally hyperbolic spacetimes, on the other hand, the Lorentzian distance function is finite and continuous just like the Riemannian distance function.

Proposition B.2.11 *Let (M,g) be a globally hyperbolic spacetime. Then the Lorentzian distance function d is finite and continuous on $M \times M$.*

Proof. See [BEE, p. 140]. □

By the definition of Lorentzian distance, a curve $\gamma \in \Omega_{pq}$ with $d(p,q) = L(\gamma)$ is the longest curve from p to q. In this case, we call γ *maximal.*

Remark B.2.2 Note that, if $\gamma : [a,b] \to M$ in Ω_{pq} is maximal, then for all s, t with $a \leq s < t \leq b$, we have $d(\gamma(s),\gamma(t)) = L(\gamma \mid [s,t])$. Motivating from this observation, we call a future-directed causal curve $\gamma : I \to M$, where I is an interval, *maximal* if $d(\gamma(s),\gamma(t)) = L(\gamma \mid [s,t])$, where s, $t \in I$ with $s < t$.

Corollary B.2.1 *Let (M,g) be a spacetime. If $\gamma \in \Omega_{pq}$ is maximal then γ can be reparametrized to be a (smooth) geodesic from p to q.*

Proof. See [BEE, p. 147]. □

Remark B.2.3 Note that if $\gamma : I \to M$ is a maximal future-directed causal curve, then by Corollary B.2.1, γ is a geodesic. Also an inextendable maximal future-directed causal (resp., timelike) geodesic of a spacetime is called a *causal* (resp., *timelike*) *line.*

A spacetime (M,g) is called *timelike geodesically complete* if every inextendable timelike geodesic of (M,g) has domain $\mathbf{R}$. But even in timelike geodesically complete spacetimes, maximal timelike geodesic connectibility may fail. But global hyperbolicity is a reasonably sufficient condition for maximal causal geodesic connectibility.

Proposition B.2.12 *Let (M,g) be a globally hyperbolic spacetime. Then, for p, $q \in M$ with $p < q$, there exists a maximal causal geodesic γ from p to q.*

Proof. See [O, p. 411]. □

Next we define the Lorentzian distance from a subset H to a point p of a spacetime.

Definition B.2.4 Let (M, g) be a spacetime and $H \subseteq M$, $p \in M$. Then the *Lorentzian distance* $d(H, p)$ *from* H *to* p is defined by

$$d(H,p) = \sup \ \{d(q,p)/q \in H\}.$$

Note that, if $\gamma : [a, b] \to M$ is a piecewise smooth future-directed causal curve from $\gamma(a) \in H$ to p with $d(H, p) = L(\gamma)$ then since

$$L(\gamma) = d(H,p) \geq d(\gamma(a),p) \geq L(\gamma),$$

γ is a (smooth) maximal future-directed causal geodesic by Corollary B.2.1.

Proposition B.2.13 *Let H be a spacelike hypersurface in a spacetime (M, g). Assume $\gamma : [a, b] \to M$ is a future-directed timelike curve from H to the point $p = \gamma(b) \notin H$ with $d(H, p) = L(\gamma)$. Then γ is a timelike geodesic and γ is orthogonal to H at $\gamma(a)$.*

Proof. See [BEE, p. 452]. □

Bibliography

[AK] H. Arikan, D. N. Kupeli, Some Sufficient Conditions For a Map To Be Harmonic, preprint.

[AMR] R. Abraham, J. E. Marsden, T. Ratiu, *Manifolds, Tensor Analysis, and Applications*, (Second Edition), Springer-Verlag, New York, 1988.

[B] J. K. Beem, Globally Hyperbolic Space-times Which Are Timelike Cauchy Complete, *Gen. Rel. Grav.* **7** (1976), 339–344.

[BEE] J. K. Beem, P. E. Ehrlich, K. L. Easley, *Global Lorentzian Geometry,* (Second Edition), Marcel Dekker, New York, 1996.

[BEMG] J. K. Beem, P. E. Ehrlich, S. Markvorsen, G. J. Galloway, Decomposition Theorems for Lorentzian Manifolds With Nonpositive Curvature, *J. Diff. Geometry* **22** (1985), 29 – 42.

[BJ] T. Bröcker, K. Jänich, *Introduction to Differential Topology*, Cambridge Univ. Press, Cambridge, 1982.

[CG] J. Cheeger, D. Gromoll, The Splitting Theorem for Manifolds of Nonnegative Ricci Curvature, *J. Diff. Geometry* **6** (1972), 119 – 128.

[DB] K. L. Duggal, A. Bejancu, *Lightlike Submanifolds of Semi-Riemannian Manifolds and Applications*, Kluwer Acad. Publ., Dordrecht, 1996.

[EL] J. Eells, L. Lemaire, *Two Reports on Harmonic Maps*, World Scientific, Singapore, 1995.

[E] J. H. Eschenburg, The Splitting Theorem for Spacetimes With Strong Energy Condition, *J. Diff. Geometry* **27** (1988), 477 – 491.

[F1] A. E. Fischer, Riemannian Maps Between Riemannian Manifolds, *Contemporary Math.* **132** (1992), 331 – 336.

[F2] A. E. Fischer, Riemannian Submersions and The Regular Interval Theorem of Morse Theory, *Ann. Global Anal. Geom.* **14** (1996), 263 – 300.

[G] G. J. Galloway, The Lorentzian Splitting Theorem Without Completeness Assumption, *J. Diff. Geometry* **29** (1989), 373 - 387.

[GK1] E. García–Río, D. N. Kupeli, Singularity Versus Splitting Theorems for Stably Causal Spacetimes, *Ann. Global Anal. Geom.* **14** (1996), 301 - 312.

[GK2] E. García–Río, D. N. Kupeli, On Affine Riemannian Maps, *Arch. Math. (Basel)* **71** (1998), 71–79.

[GK3] E. García–Río, D. N. Kupeli, Some Splitting Theorems for Stably Causal Spacetimes, *Gen. Rel. Grav.* **30** (1998), 35–44.

[GK4] E. García–Río, D. N. Kupeli, Divergence Theorems in Semi-Riemannian Geometry, *Proc. Workshop Recent Topics in Diff. Geometry.* (L.A. Cordero, E. García–Río, eds.), *Publ. Dep. Geometría y Topología*, Univ. Santiago de Compostela, **89** (1998), 83–93.

[Ha] S. W. Hawking, The Existence of Cosmic Time Functions, *Proc. Roy. Soc. London* **A 308** (1968), 433 - 435.

[He] R. Herman, A Sufficient Condition That a Mapping of Riemannian Manifolds Be a Fiber Bundle, *Proc. Amer. Math. Soc.* **11** (1960), 236–242.

[HE] S. W. Hawking, G. F. R. Ellis, *The Large Scale Structure of Space-time,* Cambridge Univ. Press, Cambridge, 1973.

[HK] K. Hoffman, R. Kunze, *Linear Algebra,* (Second Edition). Prentice-Hall, Englewood Cliffs, New Jersey, 1971.

[K1] D. N. Kupeli, The Eikonal Equation of An Indefinite Metric, *Acta Appl. Math.,* **40** (1995), 245 - 253.

[K2] D. N. Kupeli, *Singular Semi-Riemannian Geometry,* Kluwer Acad. Publ., Dordrecht, 1996.

[K3] D. N. Kupeli, On Semi-Riemannian Submersions, *New Developments in Differential Geometry, Budapest* 1996. J. Szenthe (ed.) Kluwer Acad. Publ., Dordrecth, 1988, 203–212.

[K4] D. N. Kupeli, Semi-Riemannian Transversal Maps, *Demonstratio Math.*, **31** (1998), 345 - 354.

[K5] D. N. Kupeli, On Extremal Semi-Riemannian Submanifolds, *Rend. Circ. Mat. Palermo* (2) **45** (1996), 211–232.

[K6] D. N. Kupeli, The Method of Separation of Variables for Laplace-Beltrami Equation in Semi-Riemannian Geometry, *New Developments in Differential Geometry* (L. Tamassy, J. Szenthe, eds.), Kluwer Acad. Publ., Dordrecht, 1996, 279–290.

[Ne] R. P. A. C. Newman, A Proof of The Splitting Conjecture of S.T. Yau, *J. Diff. Geometry* **31** (1990), 163 – 184.

[No] K. Nomizu, *Fundamentals of Linear Algebra*, McGraw-Hill, New York, 1966.

[O] B. O'Neill, *Semi-Riemannian Geometry with Applications to Relativity*, Academic Press, New York, 1983.

[Pe] R. Penrose, *Techniques of Differential Topology in Relativity*, Regional Conference Series in Applied Mathematics, SIAM **7**, Philadelphia, 1972.

[Po] W. A. Poor, *Differential Geometric Structures*, McGraw-Hill, New York, 1981.

[SW] R. K. Sachs, H. Wu, *General Relativity for Mathematicians*, Springer-Verlag, New York, 1977.

[Ü] B. Ünal, Divergence Theorems in Semi-Riemannian Geometry, *Acta Appl. Math.* **40** (1995), 173–178.

[V] J. Vilms, Totally Geodesic Maps, *J. Diff. Geometry* **4** (1970), 73–79.

[YI] K. Yano, S. Ishihara, Harmonic and Relatively Affine Mappings, *J. Diff. Geometry* **10** (1975), 501–509.

Index

MIX
Papier aus verantwortungsvollen Quellen
Paper from responsible sources
FSC® C105338

If you have any concerns about our products,
you can contact us on
ProductSafety@springernature.com

In case Publisher is established outside the EU,
the EU authorized representative is:
Springer Nature Customer Service Center GmbH
Europaplatz 3, 69115 Heidelberg, Germany

Printed by Libri Plureos GmbH
in Hamburg, Germany